Investigating Classroom Myths through Research on Teaching and Learning

ACS SYMPOSIUM SERIES **1074**

Investigating Classroom Myths through Research on Teaching and Learning

Diane M. Bunce, Editor
Department of Chemistry
The Catholic University of America
Washington, DC

Sponsored by the
ACS Division of Chemical Education

American Chemical Society, Washington, DC

Distributed in print by Oxford University Press, Inc.

Library of Congress Cataloging-in-Publication Data

Library of Congress Cataloging-in-Publication Data

Investigating classroom myths through research on teaching and learning / Diane M. Bunce, editor ; sponsored by the ACS Division of Chemical Education.
p. cm. -- (ACS symposium series ; 1074)
Includes bibliographical references and index.
ISBN 978-0-8412-2659-3 (alk. paper)
1. Chemistry--Study and teaching--Research--Congresses. I. Bunce, Diane M. II. American Chemical Society. Division of Chemical Education.
QD40.I65 2011
540.71--dc23
2011035065

The paper used in this publication meets the minimum requirements of American National Standard for Information Sciences—Permanence of Paper for Printed Library Materials, ANSI Z39.48n1984.

Distributed in print by Oxford University Press, Inc.

PRINTED IN THE UNITED STATES OF AMERICA

Foreword

The ACS Symposium Series was first published in 1974 to provide a mechanism for publishing symposia quickly in book form. The purpose of the series is to publish timely, comprehensive books developed from the ACS sponsored symposia based on current scientific research. Occasionally, books are developed from symposia sponsored by other organizations when the topic is of keen interest to the chemistry audience.

Before agreeing to publish a book, the proposed table of contents is reviewed for appropriate and comprehensive coverage and for interest to the audience. Some papers may be excluded to better focus the book; others may be added to provide comprehensiveness. When appropriate, overview or introductory chapters are added. Drafts of chapters are peer-reviewed prior to final acceptance or rejection, and manuscripts are prepared in camera-ready format.

As a rule, only original research papers and original review papers are included in the volumes. Verbatim reproductions of previous published papers are not accepted.

ACS Books Department

Contents

Indexes

Chapter 1

Myths: What Are They and How Can We Challenge Them?

Diane M. Bunce*

Chemistry Department, The Catholic University of America, Washington, DC 20064
***E-mail: Bunce@cua.edu**

Myths are cultural artifacts that result from our beliefs of how teaching and learning take place and interact with each other. Myths easily become "accepted truths" because they provide simple answers that are consistent with our beliefs about the teaching and learning processes. In order to challenge myths, we must be aware of the beliefs we hold and strive to provide targeted data to challenge these simple answers to multi-faceted questions. Myths can only be dispelled with convincing data and experiments that confront myths must therefore have clear research questions, suitable methodologies, and appropriate data collection and analysis. Such research must also have a firm theoretical framework so that the research results can be used to address our overarching beliefs about teaching and learning.

Introduction

Comments are often made by teachers, administrators and students alike regarding what constitutes good teaching, what problems exist in specific courses, and why learning is difficult in these courses. These comments are repeated so often that we as a community start to believe them, regardless of whether there are data available to support such beliefs. This can be the origin of unchallenged, but commonly accepted, myths. The myths seem to provide believable answers to our questions. These answers are satisfying because they are consistent with what we believe to be true about the processes of teaching and learning.

So what are our beliefs about teaching and learning? What paradigm are they based on? To illustrate the connection between our beliefs and how these belief systems influence the myths we find satisfying, let's consider two examples. These examples deal with how learning and teaching interact and the teacher's role in the classroom.

If we believe that a teacher's role as expert is to provide knowledge to students, then the myth that may seem most satisfying to explain why some students don't learn in our courses is that the students are not trying hard enough. On the other hand, if we know that they are working hard, then a secondary explanation might be that they were not prepared well enough to handle the rigors of the course we are teaching. Our belief in how teaching should take place shapes our answers to the question of why students are not learning.

If, however, we believe that students must integrate the knowledge they hold in long term memory with the new knowledge to which they are being exposed, then we might not believe that lecture or a teacher's presentation of knowledge is the most effective way to teach. In this situation, the answer to the question of why some students are not learning might not be that students are not trying or are not prepared. Instead, a more satisfying answer may be that the instructional method (presentation of knowledge by an expert) doesn't match how learning takes place.

In both of these examples, the problem is the same—some students are not learning. The answer or myth we accept to explain this dilemma is based upon our belief that either the students are not trying or the instructional method is not adequate. Both myths are equally unsupported by data. They are myths because they are cultural artifacts arising from our belief system. They are not answers based upon research into the problem. One of the dangers of myths is that they provide easy answers that may or may not be accurate. If we accept these myths because they are consistent with our beliefs about teaching and learning, we may not see the need to investigate the reasons for the problem and thus miss an opportunity to supply data to address the question and help us formulate an objective and proven explanation of the problem.

Myths are dangerous not only because they provide easy answers to problems in teaching and learning but also because they can become lodged in our collective psyche as simplistic answers to complicated problems. If we are not skeptical of these easy answers, then our research may be designed incorrectly because it may be based on a false foundation. This could result in incorrect questions being asked and inaccurate conclusions being reached.

What If?

What if we decided to challenge some well established myths? We would need data. This data should come from research that was tightly designed and executed. Such research, like all research, should be skeptical of easy answers to complicated questions. The research methodology should account for important variables in teaching and learning, and control those that cannot be measured directly. The data analysis should be a deep evaluation of the variables themselves, and would need to go beyond reporting of percentages and calculations of simple t-tests.

Investigating such research questions might require unique methodologies that include more than the usual pre/post achievement measures and survey instruments or use these instruments in new ways. Tools used in such experiments would need to precisely measure the variables of interest and may involve novel approaches or techniques to accomplish this effectively. And lastly, an adequate amount of data should be collected to allow accurate conclusions to be drawn. Lack of sufficient data could result in a false interpretation thus creating a new myth.

Myth-challenging research may need to start small, investigating myths that directly impact our classroom experience. Subsequent research could move to more global questions related to how learning takes place in general. By examining what we currently believe, and challenging ourselves to ask questions that confront these beliefs, we could develop research questions that have the potential to substantially increase our understanding of teaching and learning. The end result could be data that challenge the myths, causing us to confront our beliefs of how people learn and the role of teaching in that process. There is also a place for educational and cognitive psychological theories in this research. Without theories to help explain and provide the framework for such research, the resulting investigations might result in a "million points of light" about teaching and learning that do not coalesce to help us understand the bigger picture. Theory-based research helps build on what was learned in previous research and extends our multi-faceted understanding of the teaching and learning interaction in a more documented fashion. A more in-depth discussion of the importance of using a theoretical framework to guide research is found in the Nuts and Bolts of Chemical Education Research symposium volume (*1*). Results from individual experiments that might not make sense on their own can be tied to the theoretical framework and start to fill in the puzzle of our understanding

Purpose of This Book

The purpose of this book is to provide examples of research that address a wide array of cultural beliefs about teaching and learning manifested as easy answers (myths) to common problems. The myths range from how long students pay attention in class to which chemistry topics are more difficult to learn than others. The methods employed in these investigations include familiar tools such as surveys and achievement measures as well as newer electronic means of collecting data such as the use of clickers. It is not always necessary to formulate complicated experiments to investigate myths. Sometimes it involves asking the right question at the right time of the right person. However not all myths are incorrect. The book highlights some of these investigative approaches, both conventional and unconventional.

How To Use This Book

This book can be used in several ways. First, the book can be used to provide data to challenge some commonly believed myths about teaching and learning. Some of the research presented proves, disproves or increases our understanding of commonly held myths. Second, the chapters of this book can be seen as examples of chemical education research collected in one place so that the breadth and depth of such investigations can be compared. Some investigations look at myths in detail while others focus more broadly on several interacting variables. Both types of research add to our understanding of different kinds of myths. Third, the book showcases a range of research questions and methodologies used to investigate them. This range emphasizes the importance of matching the methodologies used to the research questions asked. Lastly, the book offers examples of research that applies the chemical education research principles presented in the previous *Nuts and Bolts of Chemical Education Research* (*2*) symposium series volume, including the importance of constructing a good research question, choosing an appropriate theoretical framework and drawing conclusions based upon the data collected.

The book is organized into three sections that include the following:

Table 1. Book Organization

Section I	Chapters 2-5	Research dealing with myths or beliefs that teachers experience in the courses they teach.
Section II	Chapters 6-9	Research that investigates more global myths that deal with types of courses, longitudinal change, or larger institutional units.
Section III	Chapters 10-12	Research addressing the use of new statistical or research methodologies and tools to investigate familiar questions

The end result of using this book is to see how carefully crafted research questions and well designed studies can add to our understanding of how learning takes place and the role of teaching in that process.

References

1. Abraham, M. R. Importance of a Theoretical Framework for Research. In *Nuts and Bolts of Chemical Education Research*; Bunce, D. M., Cole, R. S., Eds.; ACS Symposium Series 976; American Chemical Society: Washington, DC, 2008; pp 47–66.
2. Bunce, D. M.,Cole, R. S., Eds. *Nuts and Bolts of Chemical Education Research*; ACS Symposium Series 976; American Chemical Society: Washington, DC, 2008.

Chapter 2

How Long Do Students Retain Knowledge after Taking a General Chemistry Test?

Diane M. Bunce*,[1] and Jessica R. VandenPlas[2]

[1]Chemistry Department, The Catholic University of America, Washington, DC 20064
[2]Department of Chemistry, Grand Valley State University, Allendale, MI 49401
***E-mail: Bunce@cua.edu**

Teachers often report that students forget chemistry soon after a test is completed. Is this true? Is it equally true for different general chemistry courses? This chapter describes how an experiment was designed to explore this question and the research done to address it. The methodological decisions, results, and interpretation of the results are presented here. The authors describe both the research done to address the question and the decisions made that shaped the research but did not necessarily appear in the final manuscript. The results include the fact that not all students experience a decay in knowledge following a test and when such a decay does occur, it happens within the first 48 hours and remains unchanged for up to 2 weeks. Some possible explanations of why some students experience a decay in knowledge might be due to the fact that they have had fewer opportunities to test their knowledge in between tests or did not have a comprehensive final exam scheduled for their course. In addition they may have been taught using a spiral curriculum but as novices, they may not have have understood the implicit spiral nature of their curricula. Lack of student motivation is another possible explanation.

Introduction

Both new and experienced teachers are often perplexed by what they see as students instantly forgetting knowledge that they previously answered correctly on recent tests. Experienced teachers can attribute this to students' poor test taking skills and use of surface knowledge learning techniques such as memorization as opposed to deep conceptual learning. Surface knowledge can be prone to knowledge decay within a short period of time because this knowledge is not well integrated in long term memory. Novice teachers may be more inclined to blame themselves for not being able to substantially affect student learning. It is quite possible that some of each of these is true—inadequate learning strategies on the part of the student and a mismatch between how we teach and the way students learn. But there are other variables that can affect the decay of knowledge. Among them are student motivation to truly learn the concepts; course structure that supports either deep or surface learning; student aptitude and the possible mismatch between student learning and the teaching level used in the course; or cognitive overload due to how the material is presented. In summary, student decay of knowledge may find its cause in either the student, the teaching or a combination of both. Before such causes can be fully investigated, it is important to document whether the belief that students quickly forget the knowledge they have previously demonstrated successfully on recent tests is truth or myth.

In some of our past unpublished research into learning, we have seen evidence of student decay of knowledge firsthand. In one study conducted in the early 1990's on the effect of achievement in high school chemistry, when an explicit problem solving approach was taught to a treatment group, the control group scored significantly lower on an unannounced quiz on the same material they had been tested on just 5 school days earlier. Students taught the explicit problem solving method did not demonstrate a significant drop in achievement on the unannounced quiz. In a current research project testing the effect of Process Orientated Guided Inquiry (POGIL), we interviewed students of experienced POGIL teachers at three schools around the country. During these one-on-one interviews, we showed students a question that had appeared on a chemistry test they had completed within one week prior to the interview. It was discouraging to see how many students were not able to answer the question during the interview when they had successfully answered the same question within the past week. Some students even admitted that they had understood the answer to the question at one point but couldn't recall during the interview either the answer or the logic they used to solve the question one week earlier. Even in our own teaching, it sometimes surprises us how often students don't seem to remember what they had just been tested on. Although this is unverified anecdotal information from small uncontrolled samples, the feeling persists that students "forget" the chemistry they once were successful in answering. It was this anecdotal information along with reports from the literature that led us to investigate whether students truly did "forget" or experience a decay in their chemistry knowledge within a short period of time following a test. We wanted to know if this is a misperception on the part of teachers (a myth in the teaching experience) or a real phenomenon that has its roots in either the cognition or sociology of student learning or the decisions that teachers make that shape their teaching.

What Does the Literature Say?

Decay of Knowledge

Wheeler et al. (*1*) studied the differential rates of forgetting for multiple study sessions versus multiple testing occasions. The results of this experiment show that although additional studying increases the amount of recall initially, after 48 hours, the amount of retention is equivalent whether a subject uses additional studying or testing situations to help retention. After one week, the amount of forgotten material was substantially less for those who experienced repeated tests. It is the time intervals used in Wheeler et al's study that are important for our study. Based upon the results of Wheeler et al's study, we chose a 48 hour window and the delayed window of 7 days to investigate both short and long term forgetting rates. Our goal was different from that of Wheeler et al. We were interested in knowing two things, namely, 1) Is there a decay in student knowledge? and 2) If so, at what time interval after a test does it appear? We were also interested in whether a documented decay of knowledge was stable over time. In other words, if it did exist, did the decay of knowledge remain the same over an extended period or did it change (worsen or improve) over a period of up to two weeks following a test?

Hockley (*2*) makes a distinction between the decay of knowledge that occurs with item information vs. associative information. Item information is defined as individual items and associate information is the connection between events. According to the psychologists these two types of information are encoded and stored differently and thus recalled differently as well. The results of Hockley's research show that the knowledge decay of associative information is much less than that of item information recognition. Karpicke and Roediger (*3*) further investigated the variables affecting forgetting by comparing the effects of multiple testing to multiple studying opportunities. The results show that of the two, multiple testing is the critical factor for promoting long-term recall and thus a smaller decay of knowledge. This result is explained in terms of the fact that the repeated retrieval from long term memory that is necessary during testing promotes retention more effectively than additional studying occasions that do not require retrieval from long term memory.

Many of the studies dealing with forgetting or decay of knowledge including Hockley's are conducted within a psychological laboratory situation and deal with students being taught and then being asked to recall a series of nonsense syllables, words, letters or numbers. Such psychological experimentation on the parameters of learning is necessary to learn what the brain is capable of. But laboratory testing of how learning (and forgetting) takes place is not enough. Classroom research in authentic learning environments is needed to apply these psychological findings within the multivariable reality of a classroom with real students and real teachers. This means that other variables that could affect the outcome might be unknown and/or uncontrolled. The challenge of authentic learning environment research is large but the potential benefits could offer some practical guidance for how teaching and learning can optimally be combined. Arzi et al. (*4*) attempted to address this problem in their long term retention study of science learning in a school setting. Their study looked at the amount of forgetting (decay of knowledge) that students experienced over a two year period. The researchers document a significant loss of knowledge over this period but show a difference in the forgetting patterns of average/good students vs. poor students. Average or good students demonstrated an irreversible loss of a smaller part of their knowledge than did the poorer students. This seems reasonable. If students understand what they are learning, then they are more likely to be able to build connections to that knowledge in memory and thus more easily access it when needed. This is probably one of the differences between average/good students who have a deeper understanding of the new knowledge that they are exposed to than poorer students. Thus average/good students are better able to retain the new knowledge that is integrated with that already in their long term memory.

Much of the research we have about the decay of knowledge in the learning process comes from the psychology literature. But many teachers have anecdotal evidence to support the idea that students don't seem to remember information, specifically chemical information that they previously demonstrated competence in just days prior. Little experimental research has been done in authentic learning environments to document how long after a testing occasion this decay of knowledge occurs. Many authors who write books on good teaching practice attempt to bridge the gap between the psychological literature and authentic classroom environments about the decay of knowledge. They offer advice to teachers on possible causes for the decay and ways to decrease it through good teaching practices. Sousa (*5*) suggests that teachers take advantage of what we already know about the decay of knowledge, i.e., knowledge that is not rehearsed or retrieved will be forgotten and he suggests to teachers that they purposely revisit important concepts throughout a student's school experience to provide this rehearsal/retrieval. Bligh (*6*) suggests that people forget things that are not important to them and offers teachers ways to help make information more meaningful to students. Bligh also suggests that motivation is important because highly motivated people work carefully and recognize the importance of new material. Alternatively, he suggests that students who are motivated spend more time on task. This leads us to a discussion of what variables might affect the decay of knowledge in authentic learning environments.

There are several reasons why people experience a decay of knowledge after they have initially learned something. Decay of knowledge does not deal with the variables involved in the initial learning but rather the reasons why students experience a decay of that knowledge after they have demonstrated achievement of it on tests. These decay of knowledge variables include student aptitude, course structure, curricula approach, teaching style, and student motivation among others. The purpose of the research reported here was not to investigate *why* knowledge decay occurred but rather *if* it occurred. An investigation of the reasons for any observed decay should be the focus of future studies. However, a discussion of the scope of the variables that might affect the decay of knowledge is within the scope of this chapter and will be included.

Variables That Might Affect the Decay of Knowledge

Student Aptitude

Arzi et al. (*4*) report that average/good students retain more of their initially learned knowledge than their less able counterparts. The categorization of student s as good, average, or poor was based upon student chemistry achievement on a test on the Periodic Table used in the study. In planning a more general study of students' decay of knowledge as was done in our study, a measure of aptitude, rather than previous achievement was sought. Although SAT scores are used in many studies as a measure of aptitude, it is sometimes difficult to obtain permission to access such scores. In addition, many schools now accept either ACT or SAT scores and a reliable conversion formula between the two must be applied to interconvert the scores on these two tests. Since our study included both university and high school students, there was the added problem that not all the high school students in the study had yet taken the SAT or ACT. So an alternative measure of aptitude was needed. The Group Assessment of Logical Thinking (GALT) (*7*) has been used in other studies when a quick measurement of aptitude is needed. This test administered both electronically or as a paper and pencil test, usually takes 12-20 minutes to administer and consists of 12 questions, ten of which require both a correct answer and a correct reason for that answer. The test has been modified (Bunce and VandenPlas) to replace one question that might be considered offensive by people of different sexual orientations. The GALT test is a paper and pencil Piagetian test that measures students' ability to think logically. The questions in GALT are independent of any specific content area such as chemistry. Research that we have done (*8*) shows that the GALT test measures the same variable of student learning but just not as thoroughly as the Math SAT. Thus, this relatively easy to administer pretest can be used to measure student aptitude within a research study without putting undue strain on either the students or the research timeline. Once measured, the GALT score can be used as a way of categorizing students according to the variable of logical thinking. Logical thinking is a variable that is relevant to the successful study of chemistry.

We chose to include the GALT score in our analysis so that we could investigate whether the decay of knowledge was dependent on student aptitude as measured by this test of logical thinking. We were interested in knowing whether any observed decay of learning was more evident for low, medium, or high GALT students. The categorization of low, medium, and high GALT students was based upon the distribution of GALT scores for students in this study.

Course Structure

Course structure can either positively or negatively impact a student's engagement in a course. The structure is usually presented in the course syllabus and includes features of the course such as tests, quizzes, homework, class participation and other course aspects that will affect a student's grade. Course structure also includes a description of the type of feedback a student can expect in the course. It includes both features of the course and the schedule with which they will be implemented. For instance, how many full period tests will a course have? Is there is only a midterm and a final? In the case of infrequent testing, students may not fully engage with the course content until just before the scheduled tests. With 4 tests and a final exam in a 15-week semester, students must engage at least 5 times during the semester (about once every 3-4 weeks) to prepare for the more frequent tests as opposed to twice a semester in the case of a midterm and final only course structure. Course structure can also include the presence and number of intervening quizzes between tests. This feature can also affect how often students engage with the course material. Graded homework is another opportunity to engage students and provide feedback. If homework, quizzes, and tests are graded and returned in a timely fashion, students can use this feedback to refine or expand their understanding of chemistry concepts. McDaniel et al. (*9*) have shown the effectiveness of quizzing in promoting learning and subsequent retention of material on tests. Short answer quizzing which requires recall was more robust in their study than the use of multiple choice questions on tests which require only recognition. Timely feedback on the correctness of quiz answers was considered to be key in the positive effect of testing on retention in the McDaniel et al. study. The McDaniel study is an example of how quizzing interspersed with testing occasions helps promote the retention of knowledge.

The use of personal response devices (clickers) in class is also part of a course structure that can be used to engage students on a much shorter interval with the material being taught. Clicker questions are essentially ConcepTest questions: short, multiple-choice questions focusing on a single topic which are conceptual, rather than algorithmic, in nature (*10*). Clicker questions are inserted into a lecture or class period and their answers are recorded electronically on the teacher's computer. Graphs can be displayed immediately following student responses to the question. These graphs illustrate the range of understanding on a question dealing with a specific topic by showing the percentage of students who selected each answer to the question. Feedback in the form of a correct answer indicator provides students with immediate, anonymous feedback on their understanding of a particular question. Clicker questions are often embedded within PowerPoint presentations and can then be uploaded to course websites along with the correct answers so that students can review both question and answer at a later date. This course structure feature provides another means by which students can be engaged with the material to be learned. It is expected that such engagement facilitated through the structure of the course could have a positive influence on the retention of knowledge.

Curricula Approach

Hockley's (*2*) work on the differential decay of item information and associative information might be seen as the underlying principle behind the move in education from traditional or linear to spiral curricula. Traditional or linear curricula present each chemistry concept in its entirety and then move to the next concept. Spiral curricula, on the other hand, present a concept with as much detail as is needed to understand the question at hand. The curricula then return to the concept again and again adding more detail each time in association with other related concepts. The spiral curricula is sometimes explained as presenting knowledge on a need-to-know basis so that the learner has the knowledge that seems the most relevant at the moment. Such an approach can foster the connections between concepts, a hallmark of associative information, and thus has the potential to diminish knowledge decay. Spiral curricula although truly integrated in the minds of the curricula authors and teachers, may not seem obvious to the students. For instance in a highly regarded curricula *Chemistry in Context* (*11*) that presents information on a need-to-know basis, chemistry concepts are taught within the global issues of clean air, water and energy. Students who experience a complete story of global warming may not realize that the chemical concepts presented in this section are the same chemistry topics revisited in the discussion of the ozone layer or clean water. In order to be effective, the revisiting of concepts within spiral curricula must be obvious to students either through the materials themselves or how they are used in the classroom.

Teaching Style

A teacher's teaching approach, whether it is student-centered or content-centered, can have an effect on how engaged a student is in the process of learning. A teaching style that is student-centered will emphasize student engagement in learning including such techniques as cooperative groups, clicker questions and resources that provide immediate feedback for students. The course structure in such situations would make it possible for the student to demonstrate mastery in several different formats. Testing would be both formative and summative with demonstration of the student's mastery of the concepts as the ultimate goal. By contrast, a content-centered approach to teaching would emphasis "coverage" of chemical concepts by the teacher through lecture or other student-passive means. Student mastery would be tested primarily in a summative method meaning that you either know it or you don't. Decay of knowledge in courses where content coverage vs. student mastery of concepts is the goal, is expected to be larger.

Student Motivation

Motivation is defined as the internal state that captures, directs and sustains student behavior in achieving goals (*12*). In the academic process, motivation to learn is more specifically defined as a student's tendency to find academic activities worthwhile. It is hypothesized that the more motivated a student is, the more engaged he/she will be in the activities that encourage learning. The deeper that learning, the less likely or slower student decay of knowledge will be. Motivation is a multi-variable construct that many have tried to deconstruct so that measurements of some of the components of motivation can be made. In the literature (*13*, *14*) five key variables of motivation have been identified including self efficacy (*15*). Self efficacy is defined as belief in one's capabilities to organize and execute procedures that will produce the desired results. Self efficacy is often measured through questionnaires that ask people what they believe about their chance to succeed in a specific goal attainment. Science self efficacy instruments typically ask students to rate their confidence in their ability to succeed in science courses. Zuscho et al. (*16*) have shown that student scores on self efficacy instruments can predict grades in college chemistry courses. By definition we would expect students who freely chose to study chemistry to be more motivated, more willing to persevere when they encounter a difficult chemistry concept and more self satisfied with their level of understanding of chemical concepts. Students who are required to take chemistry as a pre-requisite for their major or for their general distribution requirement would be expected not to be as self-motivated. It is believed that the degree of self-motivation would be a factor in the decay of knowledge experienced by students in chemistry.

Purpose of This Research Study

The purpose of this research study was to investigate whether students experienced a decay of knowledge following a test in chemistry. Rather than choose one general chemistry course, we looked at the decay of knowledge in three separate courses with three different student populations namely undergraduate nursing students, undergraduate nonscience majors and high school honors students.

In addition to investigating *if* a decay of knowledge occurred, we were also interested in knowing *when* it occurred and if it did occur, how long it lasted within a two week period following a test. To further investigate this issue, we examined whether the decay was the same for students of differing logical reasoning ability.

Our research questions for this study were as follows:

1. Do chemistry students in a variety of general chemistry courses experience a decay in their knowledge following a regularly scheduled test?
2. How is this decay affected by the variables of type of course, length of time following the test, and student logical reasoning level (GALT)?

Sample

The sample chosen for this research included students in three separate general chemistry courses. This decision was made so that a range of students with different motivations to study chemistry could be examined. Both undergraduate general chemistry courses were taught by the same instructor and experienced similar teaching styles. The textbooks, curricular approach and course structure for these two courses differed and were geared to the goals of the course. The high school general chemistry course was taught by a different teacher and used a different textbook, curricular approach and course structure that was geared to the goals of the high school course.

The undergraduate courses included two semesters of a General, Organic and Biochemistry course for nursing students that were taught simultaneously in the Fall 2007 semester. Both courses (I and II) of the General, Organic, and Biochemistry course for nursing students used the same textbook, curricular approach, course structure, and teacher. Both courses also included students who were classified as first semester freshman. Students were assigned to one of these courses based upon their scores on a chemistry placement examination given the week before classes began. When tested for logical reasoning ability (GALT), there was no significant difference in the logical reasoning ability of these two courses (nursing I and II). Based upon all this information, it was decided to combine the students in undergraduate nursing I and II into a single nursing course variable. The one semester course for nonscience majors which used the *Chemistry in Context* (ACS) curriculum was also taught during the Fall 2007 semester.

The high school honors course included in this study was the second semester of a two semester course taught in Spring 08.

Methodology

Overview

The methodology used in this experiment was to repeat two free response chemistry questions that appeared on regularly scheduled tests at three given time intervals following the test and compare student achievement on these questions across these multiple testing occasions. In order to implement this methodology, some research decisions had to be made. Some of these decisions are described here.

Selection of Questions from Test

The decision was made to select open-ended questions rather than short answer or multiple choice questions because open-ended questions have the potential to offer richer insights into students' logic used to answer questions. Small changes in that logic could more easily be examined and reflected in the achievement grade with open-ended questions. The questions selected by the researchers were questions written by the individual course instructors which had been used previously on their exams. The researchers did not impose the use of specific questions on the classroom teachers but instead used questions from the teachers' tests that matched the research criteria. These criteria included the use of open-ended chemistry questions which the teachers believed their students should be able to answer correctly if they truly understood the chemistry. Within each course, the same grader who graded the tests was used to grade the quizzes using a rubric that had been developed by the researchers. Sample questions from each course are given in Table I.

Table I. Course and corresponding conceptual sample questions (*17*)

Course	*Question*
Undergraduate nursing I	The density of water at 37°C is 0.993 g/mL (the density changes as temperature increases). The density of normal urine ranges from 1.003 to 1.030 g/mL at the same temperature. Explain why the density of urine is greater than the density of water and how this can be used to diagnose some illnesses.
Undergraduate nonscience majors	Blue colored solutions transmit blue light and absorb red light. Are blue solutions absorbing light at a higher or lower energy than they are transmitting? Explain your answer.
High School honors chemistry	Even though the oxygen demands of trout and bass are different, they can exist in the same body of water. However, if the temperature of the water in the summer gets above 23°C, the trout begin to die, but not the bass. Why is this the case?

Time Intervals for Quizzes

Ideal times for the scheduling of quizzes were suggested by the researchers for 2-3, 5-7, and 13-14 days. The intervals had to be expanded to accommodate intervening weekends, scheduling of classes and other interruptions in the schools' schedules. In addition, the decision was made not to have the same students take the quiz at each of the three time intervals following a given test. This was done to avoid both undue familiarity with the questions and mental fatigue in answering the same questions several times within a short time frame. In order to accommodate the time necessary to remind students to take the quiz and the time necessary for the quiz to be taken, the time intervals had to be extended beyond those originally suggested. The final combination of time intervals used by all students across the courses was 2-5, 6-9, and 10-17 days.

Students Assigned To Take the Quizzes at Specific Time Intervals

If all students answered the same two questions both on the tests and at all three of the delayed time intervals, both question familiarity and mental fatigue might become intervening variables. To avoid this complication, students were assigned to one of three stratified random cohorts. Each cohort took only one delayed quiz following each test. Since each undergraduate course used three tests in this study, each cohort cycled through the three possible time intervals over the course of the experiment. For example, cohort 1 took a quiz in the 2-5 day interval for test #1, the 6-9 day interval for test #2 and the 10-17 day interval for test #3. The high school honors students cycled similarly but their course offered only two tests during the experiment.

Student Cohorts

Each cohort was determined through a stratified random sampling technique. The stratification was based upon GALT scores. Each cohort contained a random sample of low, medium, and high GALT students.

Students were not aware that they had been assigned to a specific cohort to complete a quiz. In the undergraduate courses (nursing and nonscience majors) email was used to notify individual students within a given cohort that it was time to take the quiz online. Quizzes were available online for the designated cohort during the specified time frame. Students were directed to complete the quizzes without the use of textbook, notes or other people. The time spent online to complete a quiz was noted by the researchers to detect any obvious violation of the directions to take the test on one's own. Students received credit for completing the quiz regardless of the score on the quiz. No test questions were discussed in class or with individual students nor were any test answer keys made available to students until all quizzes were completed.

In the high school honors course, a paper copy of the quiz was completed in class or lab by each cohort at the appropriate time.

Pretests

All students were required to complete the Group Assessment of Logical Thinking (GALT) test before the study began. GALT scores of the students within each course were then tabulated as seen in Table II.

Table II. Mean GALT Scores by Course

Course	*N*	*Mean*	*Standard Deviation*
Undergraduate Nursing, Part I	44	6.89	1.99
Undergraduate Nursing, Part II	29	6.66	2.54
Undergraduate Nonscience Majors	44	7.45	2.52
High School Honors	38	10.18	1.41

Results

In order to address the research questions, a statistical test was required to investigate student achievement on questions from test to quiz, but also how that change was affected by variables such as length of quiz delay, student logical reasoning level, and course type. Analyzing one student's test question score against his/her own subsequent quiz score is considered a within-subjects repeated measures comparison—comparing one individual to him/herself. For this study, then, achievement score on test and quiz questions are dependent variables and the comparison of test vs. quiz achievement is a within-subjects independent variable. In addition to this comparison, we also wanted to compare achievement scores among students of differing logical reasoning ability and with different lengths of quiz delay. Comparisons among students are considered to be between-subjects independent variables. In this case logical reasoning level and length of quiz delay are between-subjects variables (see below). Because two different kinds of independent variables were used, the statistical test selected for this study was a mixed between-within subjects analysis of variance (ANOVA).

A separate mixed between-within ANOVA was used for each of the three courses (undergraduate nursing, undergraduate nonscience major, and high school honors chemistry) because the textbook, curricula, curriculum framework, teaching practices and questions on tests and quizzes used in each course differed substantially, especially between the undergraduate and high school honors courses. Each ANOVA was first able to test for a significant overall decay of student knowledge across all tests and time delayed quizzes, regardless of length of decay or student GALT score within a specific course. In this analysis we did not investigate the effect that specific content had on the decay of knowledge. Following the test for global significance, we tested for more specific results by course, length of quiz delay, and logical reasoning ability (GALT score). To do this analysis for each course, we combined the data from all test questions into a single variable called "test question". The questions used on the time-delayed quizzes were also combined for each course into a single variable called "quiz question". Since students took each quiz at one assigned time interval, the effect of quiz delay intervals was measured by comparing the students in a course at time interval 1 with those in the course at time intervals 2 and 3. Assigning students to only one delayed quiz per test was done to control for the confounding effect of test familiarization and mental fatigue that might have resulted if each student took the quiz three times (once at each time interval) following a test. This methodological decision meant that the variable of time (for the time delayed quizzes) was a *between* subjects variable, rather than a *within* subjects variable. Students were also rotated through the different time intervals for the quizzes so that a given cohort of students did not experience the same time interval for more than one test.

Before an ANOVA can be used in an analysis, several assumptions regarding the data must be tested. First, the ANOVA requires that the dependent variable(s) be measured using a continuous scale, as opposed to using categorical data (*18*). In this study, questions were graded by a rubric as discussed above, and converted to a percentage (continuous) score.

ANOVA also requires independence of scores both within treatment groups and between groups (*19*). In this study we examine independce of scores by asking questions such as: does one student's achievement score affect the score of another student or would we expect scores between two students to be correlated in some way. If a researcher suspects a violation of this assumption of independence of scores then a more stringent alpha value can be used for the significance tests (*18*) to compensate for the dependence of scores. In this study based upon the experiment itself, student scores were assumed to be independent and a standard significance level of $p<0.05$ was used.

Homogeneity of variance is another assumption that must be met in order to use an ANOVA. We assume we are taking samples from populations that have equal variances (*18*). The most common test for this assumption is the Levene test for equality of variance. A nonsignificant result for the Levene test ($p>0.05$) tells you that this assumption was not violated. In this study, we must check this assumption for each of the three ANOVAs used for the three courses. The Levene test results are nonsignificant ($p>0.05$), as required for the ANOVA, for both the nursing and high school courses. In the nonscience majors' course, however, the Levene test shows that the data violate the assumption of homogeneity of variance ($p=0.000$ for both test and quiz question scores). The literature suggests that the ANOVA statistic is relatively robust in relation to violations of this assumption and that we can address such violations by setting a more stringent level of significance (*18*). Setting a more stringent level of significance serves as an added precaution against making a Type 2 error (*18*). For the results of the ANOVA with the nonscience majors'course, a more stringent level of significance ($p<0.01$) was adopted as suggested by the literature (*18*) as opposed to the more common significance level of $p<0.05$.

Because we are using a mixed between within ANOVA to analyze data, there is one additional assumption that must be met, i.e., the assumption of homogeneity of intercorrelations. To test this assumption, we must show that the intercorrelations among the levels of the within subjects' variable (in this case , test vs. quiz achievement scores), are equal (*18*). This assumption is tested with the Box's M statistic. The Box M statistic is sensitive, so it is generally adequate to accept a probability level greater than 0.001. In this case, all courses show a Box's M probability greater than 0.001 which means that the data has not violated this assumption.

To summarize, as a result of the experimental design used in this experiment, the three mixed between-within subjects ANOVAs (one for each course) had two *between-subject* factors (quiz time interval and GALT group) and one *within-subject* factor (the repeated measures of test vs. quiz achievement). This analysis enabled an overall within-subjects comparison of test question achievement vs. quiz question achievement, as well as between-subjects comparisons of the remaining variables and interaction effects among all variables for each individual course.

Main Effect for Test vs. Quiz Question Achievement

The results of the three mixed ANOVAs provide information about the main effects of each independent variable, as well as the interaction effects among variables. A main effect shows how one single independent variable affects the dependent variable(s). As previously discussed for this study, the dependent variables were student scores on both quiz and test questions. The independent variable we were most interested in was the within-subjects repeated measures comparison of test vs. quiz achievement for each course. The overall main effect for test vs. quiz achievement tests whether student scores changed from the test to the quiz question, regardless of logical reasoning ability or quiz time interval.

The overall main effect for test vs. quiz achievement was not significant for either the nursing general chemistry course ($F_{1,407} = 2.01$, $p = 0.16$) or the high school honors course ($F_{1,113} = 0.01$, $p = 0.92$). This means that students in these courses showed no significant difference between their test and delayed quiz achievement. For the nonscience majors, on the other hand, the main effect for test vs. quiz achievement *was* significant. The significance level of this difference ($F_{1,221} = 26.77$, $p = 0.000$) is below the conservative level of 0.01 which was chosen for this study. This significant difference exhibits a medium level of statistical power (partial eta squared = 0.108) (*18*), and shows that students experience a decrease in achievement from test (M=78.2, SD=31.4) to quiz (M=65.0, SD=31.0). Achievement means for test and quiz questions for each course are given in Table III.

Table III. Test and Quiz Question Means by Course (*17*)

Course	*Test Mean (Std. Dev.)*	*Quiz Mean (Std. Dev.)*
Undergraduate nursing	66.73 (32.78)	68.80 (31.22)
Undergraduate nonscience	78.15 (31.42)	64.99 (30.96)
High School honors	80.42 (27.43)	78.10 (32.16)

The data in Table II show that the undergraduate nonscience majors' course is the only course to show a decrease in achievement scores between test and subsequent quiz questions.

Quiz Time Interval and Achievement Interaction Effect

Interaction effects, as opposed to main effects, show us the relationship between multiple independent variables, and how they can act together to influence the dependent variable. In this study, we can examine the interaction between any combination of our independent variables. One such interaction effect between quiz time interval and test vs. quiz achievement, for example, will show us if the comparison of student test question and quiz question scores is affected by the length of time between the test and quiz.

Since the nonscience majors' course was the only course that showed a significant difference between test and quiz achievement, this was the only course where the interaction effect for quiz time interval and achievement could be investigated. This interaction effect was *not* significant ($F_{2,\ 221} = 1.92,\ p = 0.15$) for this course. This means that nonscience majors performed similarly on test and quiz questions, regardless of when the quiz was taken (2-5 days, 6-9 days, or 10-17 days after the test). Since the ANOVA did show an overall significant difference between the test and delayed quiz achievement for this course, the drop in achievement must have taken place after the test but prior to the first measured time delayed quiz (2-5 days). This is interpreted as the decay of knowledge occurring within the first 2 days (48 hours) following the test.

Main and Interaction Effects with GALT Score

The last interaction effect is that of logical reasoning ability (as measured by the GALT test) on student achievement. The mixed ANOVA provided not only an analysis of the main effect of GALT score for each course, but also how the GALT score interacted with the independent variables of test vs. quiz achievement and quiz time interval. The main effect for GALT was shown to be significant in two of the three course ANOVAs. Both the nonscience students ($F_{2,221}$=5.76, p=0.004) and nursing students ($F_{2,407}$=4.89, p=0.008) showed significant differences in test and quiz question achievement when analyzed by GALT level. This is not surprising, as GALT score has been shown to correlate with success in chemistry (*8*). The high school honors students, however, showed no significant main effect for GALT score ($F_{2,113}$=0.102, p=0.750) on the dependent variables of test and quiz question achievement. This was probably due to the homogeneity of the academic ability of the high school students who were selected for the competitive honors course. The high school honors students had the highest average GALT score and lowest standard deviation of the three courses used in this study (Table I).

The interaction effects between GALT score and the other independent variables of test vs. quiz achievement and quiz time interval in all combinations were shown to be nonsignificant. This is interpreted as student logical reasoning ability having no effect on any of the previously discussed main or interaction effects. For example, the nonscience majors' significant difference in test and quiz achievement was true for *all* students, regardless of logical reasoning ability. Low, medium, and high GALT students showed the same drop in quiz scores as compared to test scores, and the same nonsignificant interaction effect with quiz time delay in this course. While we may have originally predicted no decay for high GALT students over time, or a smaller decay over time when compared to low GALT students, we instead see that the GALT score does affect the achievement score, but does not affect the amount or time of decay. For courses where no decay was seen (nursing and high school), there is an absence of decay in *all* students regardless of logical reasoning ability. High GALT students had high achievement scores on the test and kept these high scores on the quiz. Low GALT students had low achievement scores on the test, and kept these low scores on the quiz.

Discussion

According to the data presented, only one of the three groups (nonscience majors) experienced a significant decay of knowledge following a test. This decay occurred before the first time interval of 2-5 days and did not significantly change during approximately two weeks following the test. The decay of knowledge within 48 hours is consistent with the time interval for knowledge decay reported by Wheeler et al. (*1*). Since the purpose of this experiment was to investigate if a decay of knowledge occurred, the answer is yes in some general chemistry courses a decay of knowledge is detectable within 48 hours after a test. This decay is stable over the course of this experiment (up to 17 days after a test). There is no significant effect of students' logical reasoning level on this decay. High logical reasoning students did not have a different decay of knowledge that low logical reasoning students. The only significant interaction with the decay of knowledge is due to the specific course in which it is documented.

So the natural question to ask is why do the nonscience majors experience decay in knowledge when the nursing students and high school honors students do not? Although the reasons for the decay were not specifically studied in this investigation, there is some information that was collected during the study that may partially inform the discussion.

According to the literature, additional testing or quizzing opportunities have a more significant effect on student achievement than additional studying opportunities. In this study, the course structure of both the nursing chemistry and high school honors courses included intervening quizzes between tests as part of their course structure. The nonscience majors' course did not. Two of the three courses (high school honors and nonscience majors) required the submission of homework on a regular basis. The homework was graded and returned. However, the presence of graded homework assignments does not appear to have a significant effect on the decay of knowledge. In this experiment, one course that required graded homework did not experience a decay of knowledge (high school honors) and one did (nonscience majors).

Another aspect of course structure that differed among the three courses is the presence of a cumulative final exam. The two courses (nursing and high school honors) that did have a cumulative final exam did not experience decay in knowledge. The one course that did not have a final exam (nonscience majors) did have a decay in knowledge. It is possible that students knowing that they will eventually be tested on all the material covered in the course might be at least partially responsible for the absence of a decay of knowledge immediately following the test in some (nursing and high school honors) chemistry courses.

The course curricula in all three courses claimed to be spiral curricula where concepts are revisited in ever increasing detail as the course proceeds. In two of these courses (nursing and high school honors), the topics in each chapter were listed in the textbook and students could easily identify which topics they had already seen prior to studying the new material. In the nonscience majors' curriculum, the spiral curriculum may not have been as obvious to students. The model for the *Chemistry in Context* (*11*) curriculum is to present chemistry concepts within real world problems such as global warming, ozone hole, clean water, plastics or food. It is possible that students do not separate the chemistry from the issue and therefore may not realize that polymerization is important to both plastics and proteins or that different energetic forms of radiation have different effects on bonds in both global warming and the destruction of ozone.

Although we did not purposely document the teaching style of the two instructors involved in this study, the one teacher who taught both undergraduate courses had one course (nursing) that did not experience a decay of knowledge and one that did (nonscience majors).

Student motivation is another possible reason for the difference in the decay of learning among the three courses. Nursing students may be motivated to succeed in chemistry because they regard it as a gateway course to being accepted into the clinical nursing program at the end of their freshman year. High school honors students may be concerned about their grade in all courses and the effect it will have on their overall GPA for college admission. It might be logical to think that nonscience majors, who do experience a decay in knowledge, are the least motivated to learn chemistry. Many of them are taking the course to fulfill a distribution requirement for graduation. To explore this last point that the nonscience majors in this study were not highly motivated to succeed in chemistry, we analyzed their anonymous in-house evaluations given on the last day of class.

The evaluation for nonscience majors was designed by their teacher and had been used in this course for the past ten years. Many of the questions asked for the students' rating of the effectiveness of different aspects of the course such as the use of clickers in class, TA office hours, group worksheets, labs, and the posting of annotated notes on BlackBoard following each lecture. In addition to these questions were a series of Likert scale questions that asked about the students' confidence in their ability to learn and understand chemistry both before the course started and again at the completion of the course. There were also questions asking students to evaluate their enjoyment of the course and if they would consider taking a second chemistry course if they had room in their schedules. A summary of the questions selected from the course evaluation that measure such motivation sub factors as self efficacy are included in Table IV.

Based upon this post hoc analysis, it seems unfounded to assume that the decay of knowledge in the nonscience majors' course is due solely to a decreased motivation to learn chemistry.

Table IV. Student Responses to Course Evaluation (*17*)

Question on Evaluation	*Ave. Score on Likert Scale of 1→ 5 1=Agree Strongly 5= Disagree Strongly*
1. Before this course started, I was confident that I could learn and understand chemistry	2.7
2.Now that the course is over, I feel confident that I can learn chemistry	1.8
(Average change in confidence to learn chemistry from pre to post course)	-0.9 (towards agree strongly)
6. I enjoyed this class	1.6
13. I would consider taking a second chemistry course if I had room for another science/math elective	2.4
14. This course was better than I expected it would be	1.2
15. This course was worse than I expected	4.2
Free response On a scale of 1 to 5 (1= one of the three best courses you have taken and 5= one of the worst courses), how would you rate this course?	2.0

Conclusions

There are two levels of conclusions to be made in this chapter. The first deals with the results of the investigation and the second with the decisions made during research that are often not obvious to the reader of the published research manuscript.

The results of this study indicate that there is a decay of knowledge in some general chemistry courses but not all. When this decay occurs, it takes place within the first 48 hours following a test and remains stable over a time period of at least 2 weeks. There is no differential decay of knowledge detected in any of the three general chemistry courses used in this study (undergraduate nursing, undergraduate nonscience major, and high school honors) for students of different levels of logical thinking. The reasons for why this decay of knowledge occurs in one general chemistry course and not others was not the focus of this investigation, however variables such as student aptitude, course structure, curricula approach, teaching style, and student motivation may play a role. Although not studied directly, there is some evidence that course structure such as the use of intervening quizzing between tests and the presence of a cumulative exam plus a spiral curricula approach that is obvious to students may play a role in preventing a significant decay of knowledge in some courses. The influence of these variables should be considered tenuous at this point and deserving of future, controlled experimental study.

The conclusions of this research in terms of the decisions that are made in planning and executing a chemical education research project include how to design an experiment that does not allow for competing hypotheses. One of the biggest decisions we made in planning this research was to limit the scope of the experiment to whether or not a decay of knowledge occurs following tests in general chemistry courses. We believed that this question was large enough for a single research project and that subsequent investigations could look into why such a decay occurs.

In this experiment we also chose to investigate more than one general chemistry course for the presence of the decay of knowledge. By making this choice, we have a richer data set on which to base conclusions. We chose to directly measure student aptitude using the same pretest (GALT) as opposed to using SAT scores from different administrations of the SAT exam and possible conversions between ACT and SAT scores for some students. We controlled for student fatigue in answering the same test and quiz questions multiple times within a relatively short period of time by creating stratified random cohorts of students who cycled through the quiz time intervals following tests. This cycling of student cohorts during different quiz time intervals was only possible because we collected data over an extended period of time (3 of 4 tests for the undergraduate students and 2 of 3 tests for the high school honors students).We did attempt to address intervening variables that might unduly diminish our findings by analyzing additional data that at least partially addressed the question of whether the nonscience majors were not motivated to learn chemistry and therefore a decay of knowledge for this group and not the others would be predictable. Some of the research decisions were obvious to us at the start of the study while others were made during the data collection or analysis phases. Still other decisions were a result of questions raised by the reviewers when we first submitted this manuscript for publication. The conclusions regarding chemical education research that are illustrated in this chapter are that a single investigation cannot be expected to definitely answer all aspects of the research question(s) asked. Most research studies lead to more studies to investigate additional questions raised in the course of the research. The second conclusion is that research is an iterative process that builds on what comes before it and leads to new investigations that together will provide answers to the questions we ask.

References

1. Wheeler, M. A.; Ewers, M.; Buonanno, J. F. *Memory* **2003**, *11* (6), 571–580.
2. Hockley, W. E. *J. Exp. Psychol.* **1992**, *18* (6), 1321–1330.
3. Karpicke, J. D.; Roediger, H. L. *Science* **2008**, *319*, 966–968.
4. Arzi, H. J.; Ben-Zvi, R.; Ganiel, U. *Sci. Educ.* **1986**, *70* (2), 171–188.
5. Sousa, D. A. *How the Brain Learns*; Corwin Press: Thousand Oaks, CA, 2006.
6. Bligh, D. A. *What's the Use of Lectures?* Jossey Bass: San Francisco, CA, 2000.

7. Roadrangka, V.; Yeany, R. H.; Padilla, M. J. *GALT: Group Assessment of Logical Thinking*; University of Georgia: Athens, GA, 1982.
8. Bunce, D. M.; Hutchinson, K. D. *J. Chem. Educ.* **1993**, *70* (3), 183–187.
9. McDaniel, M. A.; Anderson, J. L.; Derbish, M. H.; Morrisette, N. *Eur. J. Cog. Psychol.* **2007**, *19* (4/5), 494–513.
10. Mazur, E. *Peer Instruction: A User'S Manual*; Prentic-Hall: Upper Saddle River, NJ, 1997.
11. *Chemistry in Context: Applying Chemistry to Society*, 6 ed.; American Chemical Society/ McGraw-Hill Higher Education: Dubuque, IA, 2009.
12. Glynn, S. M.; Taasoobshirazi, G.; Brickman, P. *J. Res. Sci. Teach.* **2007**, *44* (8), 1088–1107.
13. Bandura, A. *Annu. Rev. Psychol.* **2001**, *52*, 1–26.
14. Schunk, D. H. In *Self-Regulated Learning Academic Achievement: Theoretical Perspectives*, 2nd ed.; Zimmerman, B. J., Schunk, D. H., Eds.; Erlbaum: Mahwah, NJ, 2001; pp 125−151.
15. Bandura, A. *Self-Efficacy: The Exercise of Control*; Freeman: New York, 1997.
16. Zuscho, A.; Pintrich, P. R.; Coppola, B. *Int. J. Sci. Educ.* **2003**, *25* (9), 1081–1094.
17. Bunce, D. M.; VandenPlas, J. R.; Soulis, C. *J. Chem. Educ.* **2011**, *88*, 1231−1237.
18. Pallant, J. *SPSS Survival Manual*; Open University Press: Philadelphia, PA, 2001.
19. Keppel, G. *Design and Analysis: A Researcher's Handbook*; Prentice Hall: Upper Saddle River, NJ, 1991.

Chapter 3

Student-Prepared Formula Sheets in General Chemistry: Do They Help or Hinder Learning?

Jessica R. VandenPlas,[*,1] Jack Barbera,[2] Ky Mickelsen,[3] and Ashton Hjerstedt[3]

[1]Chemistry Department, Grand Valley State University, Allendale, Michigan 49401
[2]Department of Chemistry and Biochemistry, University of Northern Colorado, Greeley, Colorado 80639
[3]Department of Chemistry and Biochemistry, Northern Arizona University, Flagstaff, Arizona 86011
[*]E-mail: Jessica.Vandenplas@gmail.com.

The amount of material to be utilized by students learning chemistry is great. To this end, many instructors allow students to use "formula sheets" on quizzes and exams. Do student-prepared sheets increase knowledge retention by forcing students to evaluate the vast body of information they have studied and select only the most important pieces, aiding in information encoding? Or do students, knowing they can utilize a formula sheet of their own devising on exams, rely on these sheets as an external storage method rather than studying and learning the material? In this study the benefit of allowing students to prepare their own formula sheets was investigated. Performance on exams using student or instructor prepared sheets was compared. In addition, performance on a course final, using *only* instructor-prepared sheets, was analyzed to determine the effect of formula sheet design on long-term retention.

Introduction

In order to be successful in general chemistry, one needs to integrate a large amount of information. While instructors may stress conceptual understanding of chemistry topics (the *why* of the content), students must also use definitions, conversion factors, equations, physical constants, and problem-solving strategies. And even if all of this information is committed to long-term memory, a student must still be able to recall the information into working memory in order to solve a particular problem in chemistry. In order to ease the burden on working memory, many instructors provide students with a "formula sheet" to be used during quizzes and exams. This sheet serves as a reference for students, supplying facts or figures, equations, physical constants, etc.

Yet even when offered the seemingly beneficial tool of an instructor-written formula sheet, many students frequently request the ability to construct their own "cheat sheets." Presumably students think that by making their own formula sheet they are "getting away" with something, knowing they will be able to include extra information the instructor would not provide, such as definitions or worked examples of problems students fear they may encounter in the testing situation.

When faced with the decision of whether to allow students to create their own formula sheets, instructors may believe the myth that doing so will be detrimental to student learning. Many experienced instructors will assert that allowing students to prepare their own formula sheets will no doubt increase student performance on course exams, as it contains extraneous information that would not be provided on an instructor-prepared sheet, but that in the long run, using their own formula sheets will decrease students' long term retention by giving them an external means of storage to reference rather than committing information to long-term memory.

The literature provides us with two hypotheses with which to frame our investigation of this myth. Dorsel and Cundiff (*1*) suggest the competing dependency hypothesis and coding hypothesis. The dependency hypothesis concurs with the myth above; that is, if students know they have a formula sheet of their own making to use during testing situations, they will rely on this sheet as external storage rather than using time encoding the material to long-term memory. This does not predict if students with their own formula sheets will do better or worse than students with only instructor-prepared sheets on exams where formula sheet use is allowed. However, it *does* suggest that students who have relied on formula sheets for external storage in the past will not be be successful on exams where formula sheet use is forbidden.

Conversely, the coding hypothesis tells us that students who make and use their own formula sheets may actually be *better* at encoding this information into long-term memory than students who do not go through this process. Theoretically, choosing the most important information to include on a formula sheet from the vast body of information studied forces students to evaluate the utility of each piece of information, perhaps seeing its place in the larger body of knowledge, and even committing it to long-term memory. In this case, students who construct and use their own formula sheets will be more successful than students who use instructor-prepared sheets not only on immediate tests with the formula sheets, but even on delayed tests where formula sheets may not be allowed.

This leads us to two research questions. First, does allowing students to create their own formula sheets affect their performance on exams in a general chemistry course where their use is allowed? To answer this, performance wias compared on course exams for students using an instructor-prepared formula sheet and students using formula sheets of their own design. Second, will using instructor-prepared or student-prepared formula sheets affect long-term retention of information in general chemistry? To answer this, students who have used both kinds of formula sheets in a semester-long general chemistry course were given a cumulative final exam, during which students could only rely on an instructor-prepared formula sheet.

Methods

Two sections of a General Chemistry I (first semester general chemistry) course, taught by the same instructor, and two sections of a General Chemistry II (second semester general chemistry) course, taught by a second instructor, took part in this study. For each course, students in one section were assigned an instructor-prepared formula sheet for each course exam, while students in the second section were asked to prepare their own formula sheets for these exams. The number of students in each section are given in Table I.

Table I. Number of Students by Course and Section

Formula Sheet Type	*Gen Chem I*	*Gen Chem II*
Student-Prepared	n = 75	n = 87
Instructor-Prepared	n = 69	n = 83

Data were collected over the course of a single semester, such that students did not overlap in these courses. In order to ensure students in the two sections of each course were equivalent prior to the beginning of the intervention (use of formula sheet on exams), student performance was measured on three in-class quizzes without the use of formula sheets. Quizzes were given in each course once per week and averaged 5 questions each. Questions were written by each instructor, and included multiple-choice, calculation, or short answerquestions. Student scores were calculated based on a maximum 25 points for each quiz. Amultivariate analysis of variance (MANOVA) was used to compare section performance across the three quizzes for each course.

Instructor-prepared formula sheets contained formulas, physical constants, and conversion factors the instructor deemed pertinent to each course exam, while student-prepared sheets were allowed to contain any information the students could fit on their page. The student-prepared sheet size was limited to one half-sheet of standard 8.5” x 11” paper per exam (0.5 page for exam 1, 1.0 page for exam 2, 1.5 page for exam 3). While the exams themselves were not cumulative, the instructors felt that the cumulative nature of chemistry itself necessitated allowing students cumulative formula sheetsas the semester progressed. In addition, the instructor-prepared formula sheets became lengthier as the course material became more difficult, and allowing students extra space on their own sheets removed formula sheet length as a variable between sections.

In order to characterize the information present on both student and instructor sheets, formula sheets were qualitatively coded for the presence of facts, formulas, and example problems. The category "facts" included items such as definitions, physical constants, conversion factors, and other statements. The category "formulas" included all mathematical formulas, such as PV = nRT. Finally, the "example problems" category encompassed all worked examples, including examples from lecture, text, or homework.

To test the effect of formula sheet type (instructor- vs. student-prepared), section performance was compared on three hour-long course exams in each section. Course exams were a mix of instructor written true/false, multiple-choice, calculation, and short answer questions, and generally covered 3-4 chapters of material. In both courses, these exams were a mix of algorithmic and conceptual questions. Student scores were calculated out of 100 points for each exam, and a MANOVA was again used to compare section performance across all three exams for each course.

Finally, to investigate the long-term effects of formula sheet use in a course, students completed a cumulative final exam. This exam was given approximately two weeks after the last hour-long course exam, and included material from all previous exams. The final exams for each section included50 multiple-choice questions and were departmental standard exams, not written by the course instructors. The final exams covered the same topics as the hour exams, but no questions were duplicated from earlier exams. All students, regardless of section, were given a departmentally-prepared formula sheet, equivalent to the instructor-prepared formula sheets used on the hourly exams. An analysis of variance (ANOVA) was conducted to determine if students performed differently on this exam based upon their previous experience in the course using either instructor- or student-prepared formula sheets.

Results

As discussed above, the researchers wanted to ensure that the sections in each course were matched in terms of chemistry knowledge prior to the start of the experiment. To do this, student scores were taken on three in-class quizzes, and compared using a MANOVA. One MANOVA was run for each course, with "section" (instructor- vs. student-prepared sheet) as a between-subjects variable, and "quiz score" (for each of 3 quizzes) as the dependent variables. The results of this test showed no significant difference between the two sections of General Chemistry I ($F_{3,140} = 0.452$, $p = 0.715$) or between the two sections of General Chemistry II ($F_{3,166} = 0.406$, $p = 0.749$) for the main effect of section (instructor- vs. student-prepared sheet). This tells us that the sections in each course performed equally well on the quizzes overall, and were therefore matched prior to the start of the intervention.

After completing the hourly course exams, students were asked to turn in their formula sheets for evaluation. Results of the qualitative analysis of formula sheet content are given in Tables II and III. These results show that student-prepared sheets included roughly equal numbers of formulas as theinstructor-prepared sheets. As expected, however, the student-prepared sheets included a larger number of facts, sometimes by orders of magnitude, and example problems, which were not included on theinstructor-prepared sheets. This tells us that, in general, students preparing their own sheets have more information, in terms of facts, formulas, and worked examples, at their fingertips while taking the hourly course exams.

Table II. General Chemistry I, Mean Number of Items per Formula Sheet

Exam	*Sheet*	***Facts*** *Mean (SD)*	***Formulas*** *Mean (SD)*	***Examples*** *Mean (SD)*
Exam 1	Student	16 (12)	3 (3)	4 (5)
	Instructor	2	5	0
Exam 2	Student	10 (8)	8 (4)	6 (8)
	Instructor	2	8	0
Exam 3	Student	24 (14)	7 (4)	12 (13)
	Instructor	4	4	0

Table III. General Chemistry II, Mean Number of Items per Formula Sheet

Exam	*Sheet*	***Facts*** *Mean (SD)*	***Formulas*** *Mean (SD)*	***Examples*** *Mean (SD)*
Exam 1	Student	5 (5)	23 (10)	1 (2)
	Instructor	3	18	0
Exam 2	Student	12 (8)	12 (5)	2 (3)
	Instructor	4	13	0
Exam 3	Student	11 (7)	8 (4)	3 (3)
	Instructor	2	14	0

To investigate the impact of this additional information, a MANOVA was used for each course to compare section scores on the hourly course exams. On these exams, each course had one section of students using an instructor-prepared formula sheet, and one section using a student-prepared formula sheet. For these MANOVAs, "section" (instructor- vs. student-prepared sheet) was again used as a between-subjects variable, with "exam score" (for each of 3 hourly exams) as the dependent variables. The results of these tests showed no significant difference between the two sections of General Chemistry I ($F_{3,140} = 0.933$, $p = 0.427$) or between the two sections of General Chemistry II ($F_{3,166} = 1.495$, $p = 0.218$) for the main effect of section (instructor- vs. student-prepared sheet) on exam score. The results show that students in each course performed equally well on the hourly exams, regardless of the type of formula sheet used. Means and standard errors for each section's performance on these exams are given in Table IV.

Finally, to study the effects of formula sheet design on long-term retention, student scores on the cumulative final exam were compared. On this test, all students used an instructor-prepared formula sheet, but sections were again compared based on the type of formula sheet used throughout the semester. In this case, one ANOVA was completed for each section, using final exam score as the dependent variable, and "section" (instructor- vs. student-prepared sheet) as the independent variable. Mean and standard errors for the final exam scores are given in Table V below. Although students who had used student-prepared formula sheets during the semester averaged a few points higher on the final exam in both courses than those who had used instructor-prepared sheets, this difference was not found to be significant in either section of General Chemistry I ($F_{1,142} = 0.241$, $p = 0.624$) or General Chemistry II ($F_{1,168} = 0.068$, $p = 0.794$). This suggests that use of either formula sheet design during the semester did not help or hinder students' long-term retention or performance on the final exam.

Table IV. Student Hourly Exam Scores by Formula Sheet Type

Course	*Exam*	*Sheet*	*Mean*[a]	*Std. Error*
Gen Chem I	Exam 1	Instructor	76.7	1.8
		Student	78.0	1.8
	Exam 2	Instructor	73.4	2.1
		Student	77.0	20
	Exam 3	Instructor	68.9	2.1
		Student	68.2	2.0
Gen Chem II	Exam 1	Instructor	81.4	1.7
		Student	80.4	1.6
	Exam 2	Instructor	74.1	1.8
		Student	70.9	1.8
	Exam 3	Instructor	76.6	2.1
		Student	78.0	2.0

[a] 100 points possible

Table V. Student Final Exam Scores by Formula Sheet Type

Course	*Sheet*	*Mean*[a]	*Std. Error*
Gen Chem I	Instructor	82.2	3.2
	Student	84.4	3.1
Gen Chem II	Instructor	96.9	3.2
	Student	98.0	3.1

[a] 150 points possible.

Discussion

Overall, students did not show increased performance on course exams when given the opportunity to create and use their own formula sheets. While this has been demonstrated in other studies (*2–5*) for subjects less mathematical than chemistry, there has been pervasive "conventional wisdom" that allowing students to prepare their own formula sheets gives students an unfair advantage over students using an instructor-prepared sheet. This myth seems to be untrue, at least as far as instructor-written course exams are concerned. Although students who wrote and used their own formula sheets had much more information at their disposal during exams, these students did not perform statistically significantly different than students without access to this extraneous information.

In addition, the myth that students who use formula sheets will rely on this external storage method rather than learning the content themselves also seems to be a myth. In this case, students did not show a significant change in performance on long-term testing measures regardless of previous formula sheet experience. While it could be argued students *knew* they would not be able to use their own formula sheets on the final exams, and therefore studied differently for the final than they had for the hourly exams, the cumulative final, given only two weeks after the third hourly exam, contained a large amount of material with little time to study. It is unlikely that students would have performed as well as they did on the final exam if they had relied on their own formula sheets as a form of external storage rather than commiting the knowledge to their long term memories.

Conclusions

Although formula sheets were not shown to affect either current or long term course achievement, the use of formula sheets may affect other aspects of student performance. Future studies will further investigate what, if any, effect formula sheet design and use might have on student knowledge attainment. For example, formula sheet design may affect performance on conceptual questions differently than on algorithmic questions, a difference that could have been lost when exams were analyzed as a whole. And if formula sheets do not affect students' academic achievement, does allowing students to prepare their own formula sheets affect some other variable that makes the use of formula sheets advantageous? Student self-efficacy, test anxiety, and cognitive load are all variables that may be affected, positively or negatively, by formula sheet design. Based on the results of this study, it seems that allowing students to produce their own formula sheet does not give them an unfair advantage during use, or hurt students' long-term retention of chemistry content. If student-prepared formula sheets can be shown to affect another variable, by decreasing test anxiety for example, their use may actually be beneficial, contrary to what the myths regarding formula use may suggest.

References

1. Dorsel, T. N.; Cundiff, G. W. *J. Exp. Educ.* **1979**, *38*, 39–41.
2. Dickson, K. L.; Miller, M. D. *Teach. Psychol.* **2005**, *32*, 230–233.
3. Hindman, C. D. *Teach. Psychol.* **1980**, *7*, 166–168.
4. Trigwell, K. *Assess. Eval. Higher Educ.* **1987**, *12*, 56–65.
5. Whitley, B. E. *Coll. Student J.* **1996**, *30*, 489–493.

Chapter 4

"Content Coverage" in a Lecture Format versus Activity-Based Instruction

Maria T. Oliver-Hoyo*

Chemistry Department, North Carolina State University, Raleigh, NC 27695
***E-mail: maria_oliver@ncsu.edu.**

This chapter addresses the myth that the most time efficient way to "cover content" is lecturing to students. The efforts toward the design, development and implementation of an activity-based instructional format known as $cAcL_2$ (concept Advancement through chemistry Lab-Lecture) are highlighted and studies conducted using this format are presented in an attempt to convince the reader that content coverage does not need to be compromised in activity-based instruction. This chapter provides examples of different types of activities and highlights how these activities are grounded by the *Two-world* specific theory and the Five Principles of curricular design. Evidence provided attempts to debunk the myth by supporting our hypothesis that activity-based instruction is a viable alternative to provide a more in-depth content coverage than traditional lecture. Performance of students enrolled in $cAcL_2$ courses demonstrates that this format can promote the use of high order cognitive skills resulting in a deeper level of understanding. In addition, attitudinal studies show that the benefits derived from this format of instruction extend to the affective domain.

Introduction

One of the most frequently asked questions regarding the adoption of "active" teaching strategies revolves around the demands on class time required to teach in formats that depart from continuous lecture. This concern is usually accompanied by a deeply rooted myth that it is impossible to "cover" all the material unless it is explicitly stated by the instructor. When pressed for time, instructors tend to present as many problems and examples as possible due to the belief that it is the most effective way for students to attain learning. This was basically my teaching approach for a considerable number of years until I went to a presentation from which the following story I recall…

…A second grade science teacher was explaining to her students about the Earth and its round shape. She spent a few days showing pictures and drawing diagrams of the Earth. As a wrap-up activity the teacher asked the second graders to draw a picture of the Earth but including themselves on the drawing. This is what one student drew:

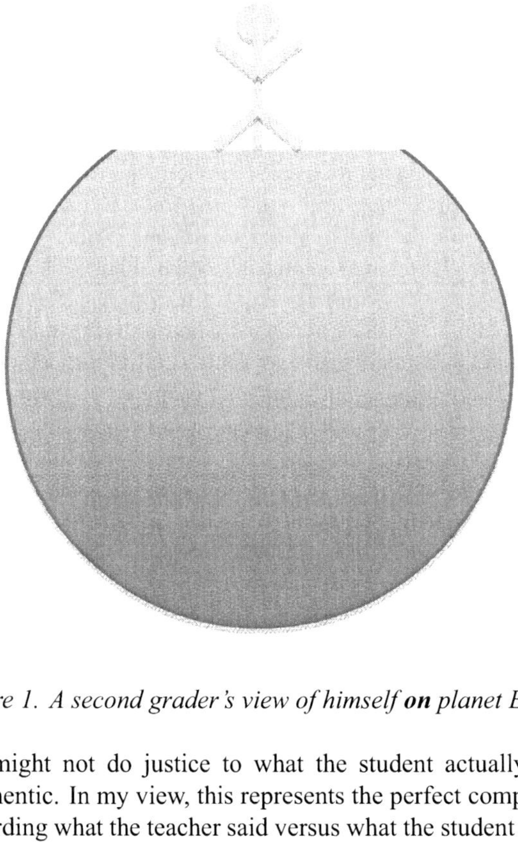

Figure 1. A second grader's view of himself ***on*** *planet Earth.*

Figure 1 might not do justice to what the student actually drew but the message is authentic. In my view, this represents the perfect compromise for that youngster regarding what the teacher said versus what the student experienced on a daily basis, as our immediate surroundings show no curvature. The realization that the same disconnect must have been happening in my classes was disturbing and this powerful image now resides in my mind. The idea that instructors' explanations will be fully captured by students and that those messages will reach the audience unfiltered and undistorted is naïve at best. Effective teaching translates into real learning and while this second-grade science teacher did her best to achieve effective teaching, her instruction did not render the expected understanding. The story has a happy ending though…

…Thinking about ways to make her second graders extrapolate content from class to real experiences she came up with an exercise. Students would place plastic insects of different sizes on top of a big beach ball and try to see themselves as the plastic creatures. What would they see? Flatness? Roundness? Why?

This was a clever intervention to make those young students think about ratio/ scale aspects that actually produced better results...so the story goes. If I were to ask the reader what made the intervention work, what would you respond? The power of such an intervention was in the format where students had to put themselves in a particular position, make observations and analyze scenarios to come up with an answer. That is the approach sought after in learner-centered strategies, which can take the shape of activity-based instruction. Time invested in making students think and reflect should produce meaningful questions and be more productive at affecting learning.

Assuming the message of this story does not resonate with the reader then a practical approach might. What if the time you spend preparing and presenting your lesson can be used differently so that students are exposed to an approach that adds value to what otherwise they could obtain from or find in textbooks and exercises? The most common question asked by instructors interested in learning more about how to put activity-based instruction into practice is:

> *How can I "cover" the material in the tight timeframe of a typical chemistry course?*

Instructors must reflect on one main issue before all other suggested paths are explored:

> *Presenting information does not necessarily translate into students' understanding.*

The introductory story in this chapter might be the perfect example to illustrate this statement, however chances are that, as an instructor, you have wondered why students' performance do not reflect your teaching efforts. The reasons might be many and varied but as academic scholars we have the resources to guide our efforts, investigate our hypotheses, document our results and shape the processes of teaching and learning. In this chapter we will take a look at a personal journey which debunk the myth that:

> **The most time efficient way to "cover content" is for the teacher to present the information in order to achieve students' learning.**

Supportive Research

A main issue to consider when addressing content coverage is the "breath versus depth" argument. This dilemma applies to both, the instructor deciding how much and how to "cover content" and the students' approach to learning. When the content coverage agenda includes a wide and extensive variety of topics, this commonly translates into exposing students to broad coverage and assessing students for low order cognitive skills since there is time to develop only superficial understandings (*1*). On the other hand, when content quantity is reduced there is more time to promote learning at a deeper level. This involves promoting the use of high order cognitive skills, which students can then apply to different contexts. Support for the "breath over depth" approach comes from the belief that students are better served by exposing them to as many relevant topics as possible. The "depth over breath" proponents are convinced that processing information is only one aspect of knowledge acquisition and that the time spent on in-depth treatment of knowledge is a more productive and effective strategy. The body of work of some educators and psychologists such as Bloom, Piaget, and Vygotsky assumed in-depth learning as the best choice to attain understanding but the "breath versus depth" argument is a long-standing debate that has provided strong points of view in both camps. However, "the educational research literature provides precious little empirical evidence that supports either approach or one over the other" (*2*). Contributing to this special void, Schwartz et al. set out to study the impact on college students' performance of the amount of content covered in their high school science courses. This comprehensive study included 8310 students enrolled in science courses (biology, chemistry, and physics) in 55 different U.S. higher education institutions. Their results showed that teaching for depth in at least one topic in high school science leads to better performance of those students in introductory college level science courses (*2*).

A valid sentiment expressed by those who develop teaching materials is that a wide gap exists between educational theories and curricular design. "General philosophical orientations to educational matters – such as constructivism – are important to educational matters, but often fail to provide guidance in organizing instruction" (*3*). This is particularly reflected in the fact that the same instructional strategies are advocated independently of the theory evoked (*4*). Some conclusions resonate strongly with curricular designers: "Even though these theories [cognitive and social theories of learning] are extremely valuable, they are still quite general and do not provide a sufficient understanding of underlying mechanisms to give much guidance for curriculum development" (*5*). The challenge is to find theoretical elements or tools that can be interwoven with the operational expression of a specific curricular design.

Confronted with these issues, Tiberghien et al. set out to propose a theoretical framework for the development of teaching resources, specifically a "teaching sequence" for physics topics (mechanics) (*6*). This major undertaking produced the *Two-World* framework (referred to as a *specific* theory), which provides two levels of knowledge categorization. The first level differentiates between theoretical and experimental levels or *worlds* where explanatory ideas are associated with the theory and models category while observations, perception, and measurements relate to objects and events. The second level further categorizes knowledge into discipline-specific knowledge (school taught physics) and everyday knowledge (life experiences). This finer distinction is strongly dependent on the context of use due to the interplay between acquired knowledge becoming everyday knowledge at some point in the learning process. "The *Two-World* specific theory guides the design process due to the necessity of differentiating between theoretical elements and objects and events" since this theory acknowledges that the relation between theory and experimentation "is not direct at all…" (*6*). Therefore, connections need to be made explicitly using tools and resources. This directs the focus of activities to primarily look at four-way interactions: relations between objects and events; relations from objects/events to theories/models; relations from theories/events to objects/events; and relations between theories and models. Figure 2 is our rendition of this framework at the first level of categorization.

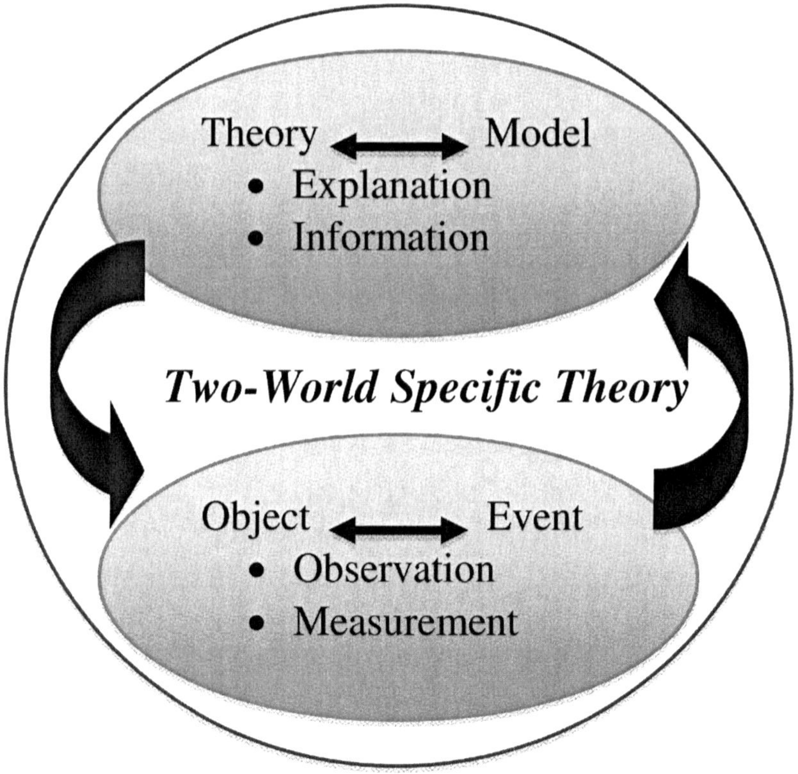

Figure 2. Diagram of the Two-World specific theory.

The design of instructional materials cannot proceed in a vacuum. The theoretical underpinnings are only one of the elements to consider. The inherent complexity of the processes of teaching and learning and the tight interactions between them demand that at a minimum the learning environment, expectations and/or learning goals, and assessment methods be attentively considered when curricular materials are developed. Meyers and Nulty extended Biggs' 3P (presage, process, product) model of learning and teaching (*7*) and provided their articulation of Five Curriculum design principles, which apply to teaching and learning materials, tasks, and experiences (*8*). These principles are included in Table 1.

Table 1. Curriculum Design Principles*

	Curriculum Design Principles
#1	impart relevancy and authenticity (real-world)
#2	promote constructive, sequential and interlinked ideas
#3	require students to use and engage with progressively higher order cognitive processes
#4	align with each other and the desired learning outcomes
#5	provide challenge, interest and motivation to learn

* Reference (*8*).

These principles provide a practical framework that combined with the *Two-World* theoretical framework set a rich foundation for the development and evaluation of activity-based instruction.

Building an Environment To Support Activity-Based Instruction with Curricular Robustness

The results of the intervention in the introductory story revealed to the teacher that there are different ways to "cover content". The educational literature identifies time constraints as the deciding factor to adopt either a superficial or an in-depth approach to teaching. As an educator, I favor depth over breath and the challenge then becomes how to use restricted class time to attain such depth. This is at the core of the myth this chapter attempts to address.

We hypothesized that activity-based instruction is a viable alternative to provide a more in-depth content coverage than traditional lecturing. We set out to study if the activity-based instructional curriculum we developed would promote the use of high order cognitive skills in students and result in better achievement as measured by students' performance. We have chosen a theoretical (*Two-World specific theory)* and a practical framework (Five Principles for curricular design) to take a closer look into the activities we designed for the $cAcL_2$, concept Advancement through chemistry Lab-Lecture, a format of instruction where hands-on activities are the preferred medium for topic explorations. This format is the chemistry dissemination project of SCALE-UP, Student-Centered Active Learning Environment for Undergraduate Programs (*9*). Time constraints are tied into the format as the topics are prescribed for the general chemistry curriculum.

We broadly describe activity-based instruction as an approach which incorporates multiple and varied ways of exposing students to content. Such practices involve both teaching strategies (such as inquiry-guided instruction and collaborative work) and available resources (simulations, live demonstrations, experiments, clickers, etc.). In our case, we emphasize a hands-on, minds-on approach and use wet-chemistry, worksheets, and Internet resources in our attempt to engage students in the process of learning. Four decades ago, Edgar Dale published the "Cone of Experience" (*10*) which has been extensively modified through the years and eventually referred to as the "Cone of Learning" or "Learning Pyramid(s)". Most versions give a correlation between retention of material and the nature of involvement for a variety of educational means instructors have available (Figure 3). Due to questionable sources for the data, Thomas Lord more recently embarked in his own experimental design and actually confirmed those correlations (*11*). What these findings tell us is that it is not until we *do* something with the information provided, that the chances for effective instruction (achievement of learning by students) increase. In activity-based instruction, we complement resources that are more passive (from the top of this triangle: reading, hearing, watching) with others that fall in the lower portion of this pyramid such as discussing, simulating, or doing the real thing. For example, demonstrations performed by the instructor may be complemented with group discussions, animations, or students' presentations of related applications.

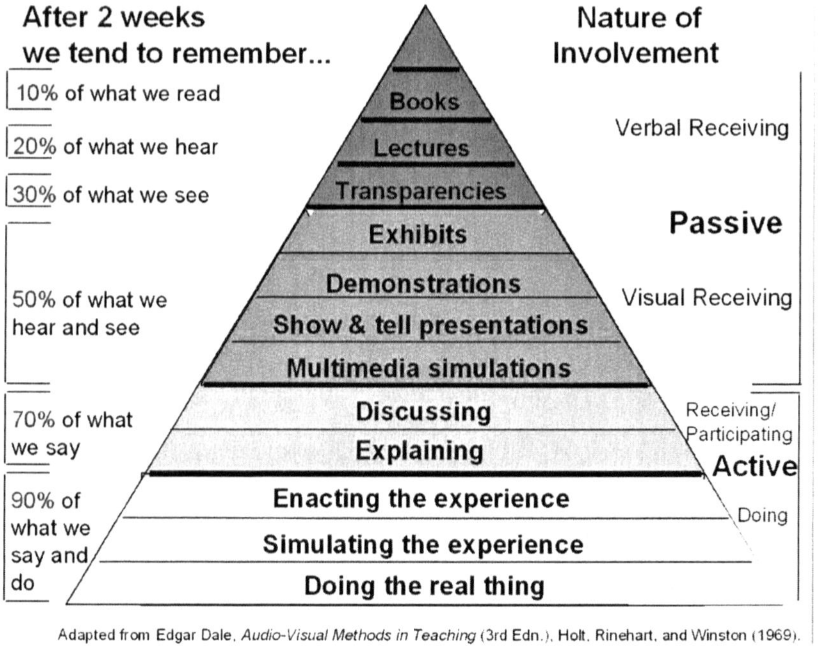

Figure 3. Modified Cone of Learning.

Inquiry-guided instruction is practiced where students are guided with targeted questions to answer unknowns and emerging questions are "answered" with questions that promote thinking critically about the activity performed (Principle #3). The key is to engage the mind of the student in order to maximize the opportunity for learning. The curricular materials have been designed to appeal to students' interest (Principles #1 & #5) with the intent that students' learning efforts become commensurate with the expected level of understanding (Principle #4). When activities become relevant (Principle #1) students tend to beengaged (Principle #5) and the chances they will dedicate more time to the topic (both in-class and outside of class) increase. We assumed that students' efforts would compensate for the restricted time in class since deeper understanding demands more time to process knowledge.

From a Practical Approach to the Generation of Fundamental Ideas

As it is evident in any educational research study, the learning environment is a key element in achieving learning outcomes. Instructors commonly choose the curricular materials but the learning environment is the result of the instructors' efforts towards teaching and the students' input into the learning experience. In order to provide instructors with some guidelines on how to choose curricular materials, I will start by describing how we developed and have used an activity-based curriculum in the $cAcL_2$ format. Since curricular materials in this format substitute for the "lecture" they must provide appropriate content. Choosing these materials is probably the most time consuming task' however, it is comparable to deciding on a new textbook or developing a new course. Whether you decide to develop your own activities or search for available resources, it is a process better done at a slow pace. In our case, we have developed about 100 activities for general chemistry that instructors may download for free (*9*). The process started by developing and testing 6 activities in one semester, then 4 more during the summertime, and eventually increasing the number of activities per week to the point where now approximately 85% of the time in class is spent working on activities. For example, in a two-hour class period a maximum of 20 minutes is used for a power lecture. Activities take a variety of forms from written and verbal communication to collection of data from wet chemistry experiments or demonstrations.

Activity efficacy refers to how well the activity incorporates the content intended to be explored, how much student engagement is involved, and what learning gains are expected versus those actually obtained. As instructors incorporate an activity multiple times in the classroom, they gain experience and can make their own determination as to the efficacy of a particular activity. When activities prevail as the main teaching resource, a more holistic approach can be taken. A practical example may be the best way to illustrate activity efficacy. The unit in quantum theory is a challenging one when it comes to incorporating demonstrations into a lecture format. Table 2 includes the sections in which this unit has been fragmented in the textbook we use (*12*) and the activities that have been substituted for the lecture mode in the classroom.

Four activities were developed to address the content material in this unit. It is worth noting that the class time reserved to "cover content" in this unit remained the same. In other words, students in the activity-based instruction were in sync with students in the lecture format. The key factor to keep in mind is that now the format for "content coverage" has changed dramatically. These activities aim at getting students physically involved and promoting that effort engages their minds as well. Each activity contains "comprehension question(s)" to assess if students have grasped the essence of the material (Principle #4). In our application of the *Two-World* framework to chemistry topics, we considered any chemical particle (micro) or substance (macro) , an object and any observable and/or measurable chemical occurrence, an event. Micro "objects" are situated in the material world using simulations or animations as intermediaries. Information obtained by our senses (hearing, touch, smell, etc.) belongs to the material world. We view models and theories as the other *world* or a conceptual recognition of knowledge. These activities are briefly described below to show the different forms activities may take.

Table 2. Topics, demonstrations, and activities that comprise the quantum theory unit

Unit: Quantum Theory		
LECTURE	*DEMONSTRATIONS*	*ACTIVITY-BASED*
The Nature of Light	Overhead projector Spectrometer Spectral tubes	Fingerprinting
Quantization	Blackbody radiation	Photoelectric effect
The Bohr Model		Seeing the light
Quantum Theory	Standing waves on a string	The atomic hotel
Electron configurations		

Fingerprinting

The fingerprinting activity (*13*) uses commercial bar codes as an analogy for students to explore how information can be extracted from a line (relation between object and event). The correlation between commercial bar codes and line emission spectra emerges from the fact that both contain information in the form of lines and those lines identify a specific product or element (relation between objects/events to theory/models). Students discuss how computerized scanners read light patterns reflected from the bar code (relation between object and event). The dark lines in the bar code absorb light, while the white spaces reflect the light. Students are asked to draw both an absorption pattern for a bar code and a predicted bar code from a reflectance pattern (relation between objects/events to theory/models). This exercise assesses student understanding of whether light is reflected or absorbed. The reflectance pattern is the key link to the line emission spectra generated by atoms and this leads to the discussion of how absorption and emission spectra are obtained (relation between theory and models).

Decoding bar codes for purposes of commercial product identification leads the discussion into spectra identification. Students are able to see that the gaps in absorption spectra correspond to the lines in emission spectra (relation between object /event to theory/model). Emission spectra not only contain information in lines like bar codes do, but they physically resemble bar codes as well. When a reflectance pattern is overlaid with its respective bar code, the peaks fill in the white gaps creating one continuous spectrum (relation between object/event to theory/model). This leads to the relationship between absorption and emission (relation between theory and models). For these exercises, students have access to Internet resources where they can observe these types of spectra for elements. The activity is concluded with a discussion of the hydrogen emission spectrum (relation between theory and models). Figure 4 depicts a diagrammatic summary of the topics this activity entails.

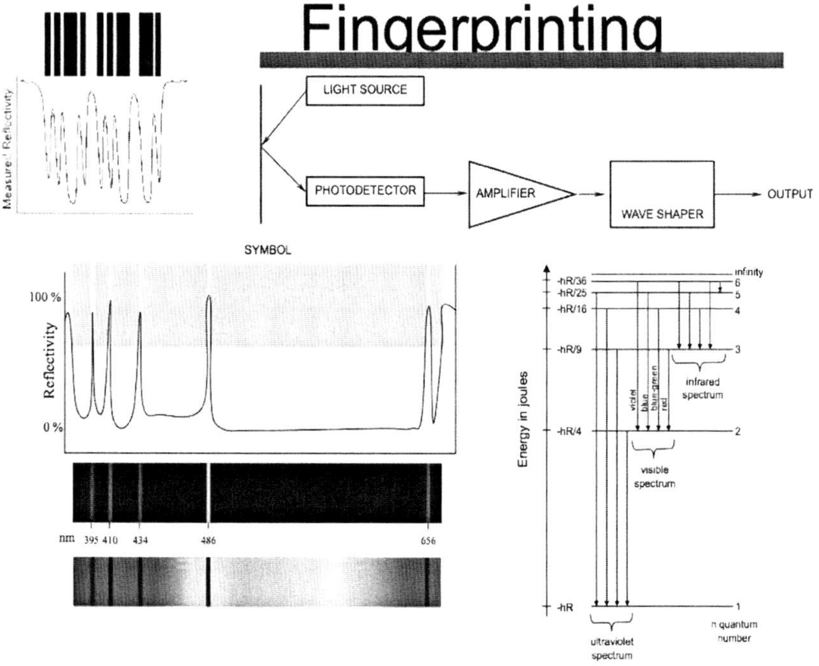

Figure 4. Fingerprinting activity relating commercial bar codes and elemental spectra.

Photoelectric Effect

In this activity, the minimum energy required to eject an electron from a metal is found for different metals using a simulation (relation between object and event). This simulation also shows the role intensity plays in this effect (relation between object/event to theory/model). In addition to the photoelectric effect, students explore the relationship between energy, wavelength, and frequency via graphical relationships (relation between theories/models and objects/events). Each student is assigned a metal and each student practices graphing skills using the data collected (relation between object and event). Figure 5 shows students' data for aluminum.

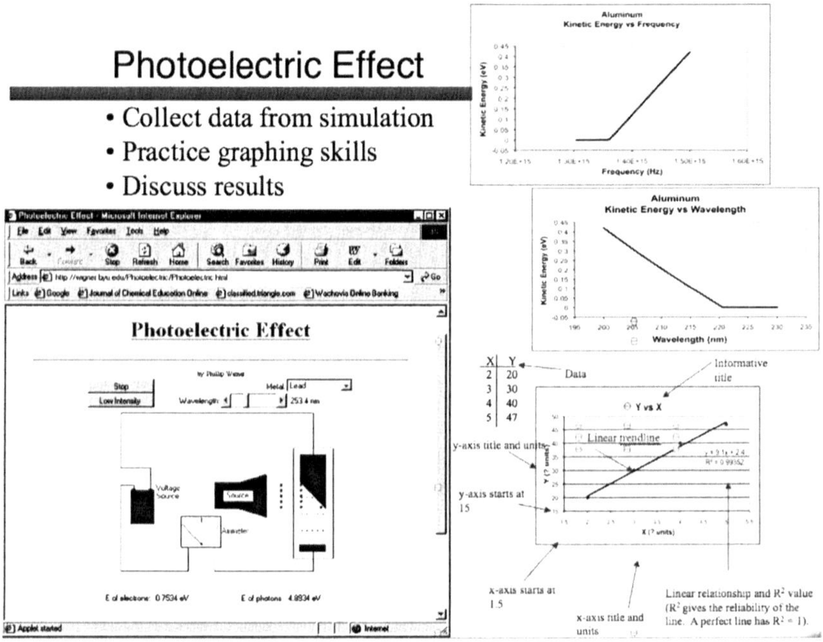

Figure 5. Simulation activity to explore photoelectric effect.

Keeping track of simulations and animations' Websites can become a time consuming task for the instructor. The particular Website depicted in Figure 4 was an excellent site used for years in our general chemistry course but it is no longer available. A quick search for "simulations on photoelectric effect" produces a number of sites that instructors may explore and choose from.

Seeing the Light

This activity consists of three parts:

- using spectroscopes to observe emission lines from different gas discharge tubes and comparing their observations to the spectra shown at a Website (relations between object and event and between objects/events to theory/models),
- placing colored solutions between the light and a diffraction grating on an overhead projector to relate observed to absorbed colors and construct a color wheel (relation between object and event),
- collecting data from a Website on additive colors (emission) to observe differences between mixing colors of light and mixing paint, and on subtractive colors (absorption) to observe the relationship between colors and white light (relation between theory/model to object/event).

Via these explorations, students study light, color, and their relationships to energy, wavelength, and spectra. Our classroom facility allows us to set stations for each part and students rotate through them to collect data. Alternate viable settings might include a laboratory with different benches for each station or assigning students to different stations and group sharing the information collected.

The Atomic Hotel

This activity simply involves working on a worksheet that uses the analogy of a hotel and its rules to relate to quantum mechanical concepts. The Atomic Hotel is a special hotel designed for electrons with strict policies on how to fill the rooms (*Aufbau principle*), room occupancy (*Hunds' Rule and Pauli Exclusion Principle*), room configurations at each level (*s,p,d,f*), identification of guests (q*uantum numbers)*, and evacuation plans (*ionic configurations)* where the exit signs are placed next to certain room configurations. After students become familiar with this special hotel and its rules (relation between objects and events), the terminology is introduced, the quantum concepts are discussed (relation between objects/events to theories/models) and electronic configurations are practiced using the periodic table (relation between theory and models).

These four activities have replaced the lecture mode for this unit and the topics are "covered" within the same timeframe as the lecture-based class. They are representative of the variety of activity modes there are including the use of Websites, simulations, worksheets, wet chemistry and demonstrations as well as real-life analogies to make chemistry more "tangible" to students. All activities demand students *to do things* and then *reflect* on what they have done.

Principles of Curriculum Design

Quantum theory is a challenging topic to make it an "authentic, real-world, and relevant" experience for students as our first principle for curriculum design recommends. However, the *Fingerprinting* activity is a popular one among students, as they show great interest in finding the actual product their commercial bar code represents which in turn motivates them to observe the spectra of particular elements. This activity brings real-life experiences into the classroom. In *Seeing the Light,* students' experiences of light, color, reflection, and diffraction are explored under a more "tangible" lens. In addition, the questions students discuss are real applications of the concepts explored. For example, one of the comprehension questions in the *Photoelectric Effect* activity is: "Explain why the photographic film can be handled in a "dark" room lighted with red light but is ruined if exposed to the same intensity of yellow light." Or in *Seeing the Light*: "Why do airplane pilots and air traffic controllers wear orange colored lenses on bright sunny days, hazy days, and foggy days?"

As can be observed, each of these activities builds from the previous one in a sequential and interlinked fashion. In *Fingerprinting,* students are exposed to emission and absorption spectra and to the importance of those "lines" which by the second activity, *Photoelectric effect,* are defined in terms of wavelength, frequency, and their relationship to energy. In *Seeing the Light,* the "new" terms and relationships are reinforced via exploration of additive and subtractive color patterns. The energy theme continues into *The Atomic Hotel* for electrons. These activities were designed to be "cumulative in their effect" with explicit assessments to monitor students' progress in their understanding of content material.

High order cognitive skills are promoted in several ways. Textbook exercises are complementary to the course work but the activities pose relevant questions, problems, and discussions framed to allow students the opportunity to think critically about options, evaluate observations, and propose explanations to the phenomena observed. Assignments are structured so that students build a foundation upon which to expand their knowledge. Formal assessments (homework, quizzes, and tests) also reinforce explicitly the learning outcomes so that students recognize the need to study more in-depth. Exams contain a subset of questions considered to be challenging and involving high order cognitive skills, specifically application and evaluation types.

Principle #4 directed the development of these activities as the process involved research on misconceptions and outlining learning objectives to address the content and the potential misconceptions. These objectives also guide the kind of questions prompted by instructors and the problems students do. This adds to the alignment requirement and provides a way for checks and balances regarding the aims and desired outcomes of the activities. Each activity has been developed using a template that contains the following sections: Title, Time (duration), Topic, Type (probe or investigation), Level (introductory, intermediate or advanced), Overview, Equipment and Materials, Objectives, Misconceptions, Other Student Difficulties, Prerequisites, Activity Table (suggested sequence of tasks, reasons for doing the tasks and practical notes), Related Activities, References, Discussion, and Supplementary Material and Comprehension Questions.

If Principle #5 (provide challenge, interest and motivation to learn) is achieved, students become engaged and tend to spend more time on learning, which in turn, compensates for the tight time on task provided by the classroom experience. This is probably the most difficult Principle to address as students come from different backgrounds, bring different skills and interests into the classroom, and change every semester. The instructor and the learning environment play a bigger role than the activities themselves to achieve this Principle. Attitudinal studies offer a resource to monitor if this has been accomplished.

The conscientious application of the Five Principles of curriculum design aims to increase the effectiveness of teaching; the engagement of students with the material in and out of the classroom; the time dedicated by students to achieve the learning objectives; and the levels of cognitive skills used and required for problem solving. After the curriculum was developed and implemented during several semesters, the research questions emerged:

- Can we show that activity-based instruction is a viable alternative to provide a more in-depth content coverage than traditional lecturing?

- Have we been able to promote in students the use of high order cognitive skills?
- Is the $cAcL_2$ format conducive to change students' attitudes toward the discipline?
- Are there specific elements in this format that promote learning better than others?

Chemical Education Studies

The way curricular materials are used influences their effectiveness in the classroom. Instructors play a major role in choosing the materials and the instructional strategies, deciding the order, type, and duration of the activities, and assuming or assigning roles for themselves and the students. Our work developed itself in phases starting with a "low-risk" incorporation of a few activities during a semester. As these were implemented, we gathered qualitative data on activities' duration, cost, availability of resources, chronology of steps, perceived relevance and engagement level of students. This information was used during the elaboration phase to direct the efforts of the development and implementation of subsequent activities. This involved a reiterative refinement process. As I became more committed to activity-based instruction and the $cAcL_2$ format, the need to document our efforts became more pressing. The evaluation phase started two years after the first activities were incorporated into a general chemistry course. Four studies were conducted that investigated the effects of this format on students' performance and attitudes (*14*, *17*, *21*, *22*). These studies represent the scholarly attempt to validate our hypothesis that activity-based instruction can provide in-depth content coverage and debunk claims that the most time efficient way to "cover content" is for the teacher to present the information.

The main goal of adopting and developing activity-based instruction was to improve student understanding as measured by students' use of high order cognitive skills, HOCS. The first study involved two sections of the same course, general chemistry I, where one section received traditional lecturing (control section) while the other was taught by the same instructor in the $cAcL_2$ format (*14*). The content of the course was exactly the same for both sections with the same detailed syllabus and the same perspective and emphasis on different topics. Special attention was paid to the time on task where the maximum allowable time for topics' activities in the $cAcL_2$ section was strictly limited by the time spent in the control section on that particular topic. The format of instruction was the main difference but students were exposed to the same content. For example, demonstrations conducted in lecture were transformed into an activity that students would perform themselves in the $cAcL_2$ section. Even though exams for both sections were very similar in length and amount of material covered per exam, they could not be identical due primarily to the fact that assessments must reflect instructional practices. In the control section, homework and examinations used predominantly multiple choice and short answer formats while the $cAcL_2$ section had problems and extensive questions. In addition, students in the control section were not exposed or involved in formal discussions as extensively as the $cAcL_2$ students. Therefore, selected groups of questions considered to be challenging but appropriate for both sections and involving HOCS were included in all exams. These subsets of questions in each exam constituted the database for the analysis of exam scores.

The equivalency of comparison groups was demonstrated taking into consideration academic major, gender, ethnicity, and previous number of chemistry courses taken by the students enrolled in both sections. A three-way ANOVA model including terms for class, major, exam, and all two-way and three-way interactions was used to analyze the data. The significant statistical differences between the two sections strongly suggest that activity-based instruction positively influenced the learning experience of students. For example, the bottom 25% of the student population in the $cAcL_2$ format outperformed the equivalent student population in the last three out of four examinations. It is also significant that the final comprehensive exam questions showed significantly higher mean and median values in the $cAcL_2$ when compared to the traditional control (lecture) section. These results encouraged us and provided the first validation to our efforts (*14*).

Due to the widely accepted assumptions that achievement and attitude are positively interdependent (*15*) and that affective variables are as important as cognitive ones in molding students' learning (*16*), we investigated if the same positive results attained in achievement would be found in students' attitudes toward the $cAcL_2$ format and the discipline (*17*). The surveys developed for this study were based on two tools, the Derived Chemistry Anxiety Rating Scale, (*18*) and a tool developed by Grinnell College and the University of California Berkley to compare a modular-teaching approach to a traditional-lecture format. Items included questions designed to monitor attitudes toward learning the discipline and perceived anxiety in learning, evaluation, and chemical handling. Pre- and post-surveys were administered electronically to students during the first and last two weeks of classes. Scores were used to compute residualized gains scores (RGS) in learning attitude (*19*) and t-tests were used to analyze changes in student anxiety. In order to eliminate a bias in attitudes toward the instructor by students in these two different formats, end-of semester departmental evaluations of the instructor were also analyzed. These reflected no significantly different perception of the instructor by students in the two sections. Positive differences in student attitudes toward learning were markedly obtained as 77% of the student population in the $cAcL_2$ section showed RGS gains above projected changes from the control section. These RGS tend to attenuate some of the difficulties associated with comparison of raw scores from complex factors affecting attitudinal responses (*19*, *20*). The RGS findings were confirmed by student responses given to open-ended questions in the departmental end-of-semester evaluation survey. The t-test analysis found that no significant difference in anxiety was shown by the $cAcL_2$ population in any of the three factors measured; learning-chemistry, chemistry-evaluation, and handling-chemicals. Since attitudes are so complex and the $cAcL_2$ format embraces a variety of elements (such as activity-based instruction, inquiry-guided and collaborative grouping approaches), we subsequently investigated these in more detail.

A second attitudinal study used triangulation to monitor changes in attitudes of students enrolled in a $cAcL_2$ second semester general chemistry course (*21*). Specifically, we monitored attitudes toward both objective (activities) and subjective constructs (educational strategies) via this qualitative cross-validation method. The triangulation scheme used data collected via different qualitative research methods that included 55 sets of interviews, 116 sets of surveys, 90 entries in students' reflective journals, and 38 field note entries. For the interview process, we followed a random stratified sampling procedure (based on major, gender, and ethnicity) to assign every student to one of three one-on-one interviews conducted during the semester. Two types of surveys were collected including a standard departmental survey and pre- and post- questionnaires developed specifically for the course. Even though triangulation was achieved only in the case of monitoring attitudes towards objective constructs (such as graphing activities), the vast and revealing data obtained from these methods strongly suggest that students become more receptive and positive toward the teaching strategies as the semester progressed.

We interpreted the results on student performance described earlier (*14*) to reflect student use of HOCS due to the fact that students' scores on subsets of challenging questions were significantly different. To further shed insight into this aspect we designed a study to show if the $cAcL_2$ format would indeed promote the use of HOCS (*22*). For this purpose, we utilized the GOAL (Gather, Organize, Analyze, and Learn) problem-solving protocol whose most distinctive aspect from other methods is the Learn step as it requires students to think about the problem beyond the answer, emphasizing the meaning of the question and the implications of the answer. The GOAL protocol is designed to encourage students to take a more active role by thinking beyond the prescribed steps required to reach the solution of a problem. The basis of the study involved data collected from four qualitative problems specifically designed to be solved with the GOAL protocol and assigned over the course of a semester. These problems served as a platform to encourage students to use HOCS in their Learn responses. Students' reflections on their solutions were characterized (not the solutions themselves) and the Learn responses were evaluated based on Bloom's Taxonomy of Cognitive Development (*23*). Four categories emerged from students' responses that represented different levels of abstract thinking. Results showed that students were increasing their abstract thinking about problems as the course progressed with a drastic drop in the number of responses at the lowest cognitive category from the first problem to subsequent problems. These findings support our claims that activity-based instruction as conducted in the $cAcL_2$ format promotes students' understanding at a deeper level.

Final Thoughts

This project has been a journey that started with a story of a second grade teacher and her efforts to make her teaching more effective. The revelation that there is so much more to teaching than good intentions, organizational skills, and content knowledge propelled me into a completely different direction from the one my formal training provided. The design of activities started with the simple goal of developing instructional materials for chemistry education. The practical approach involved taking small steps, one at a time and building upon those practices that worked well in the classroom. To turn new ideas into practice demanded a lot of personal learning, reflection, and willingness to take risks.

The examples in this chapter illustrate different types of activities while the ties to the theoretical and practical frameworks discussed here highlight how to design instructional materials to promote a deep level of understanding. The literature provided the *Two-World* specific theory and the Five Principles of curricular design to anchor the design of the activities developed. This is essential for educational improvement to grow.

Dissemination of our work continuously raised questions regarding time constraints and effectiveness of our activity-based instruction approach, $cAcL_2$. We have been successful at implementing activity-based instruction with approximately 85% of the class time dedicated to hands-on, mind-on learning; however, the popular myth that there is not enough time to cover required content via activity-based instruction persists. Anecdotal evidence we experienced was encouraging, but the need to demonstrate that activity-base instruction is a viable approach to in-depth content coverage emerged. Our studies have shown the benefits of the $cAcL_2$ format addressing both students' performance and attitudes. *Our experience supports the idea that content coverage is not compromised in activity-based instruction but transformed.* Activity-based instruction is a viable alternative to achieve in-depth content coverage. These evidence-based methods are expected to promote changes, and I certainly hope that this chemical education research journey is used to directly inform practice of education.

References

1. Newmann, F. M. *Phi Delta Kappan* **1988**, *69* (5), 345–348.
2. Schwartz, M. S.; Sadler, P. M.; Sonnert, G. *Sci. Educ.* **2009**, *93* (5), 798–826.
3. Cobb, P.; Confrey, J.; diSessa, A.; Lehrer, R.; Schauble, L. *Educational Researcher* **2003**, *32* (9), 10.
4. diSessa, A. A History of Conceptual Change Research: Threads and Fault Lines. In *Cambridge Handbook of the Learning Sciences*; Sawyer, K., Ed.; Cambridge University Press: Cambridge, 2006; pp 265−282.
5. Rea-Ramirez, M. A.; Clement, J.; Nunez-Oviedo, M. C. An Instructional Model Derived from Model Construction and Criticism Theory. In *Model Based Learning and Instruction in Science*; Clement, J., Rea-Ramirez, M. A., Eds.; Springer: New York, 2008; p 27.
6. Tiberghien, A.; Vince, J.; Gaidioz, P. *Int. J. Sci. Educ.* **2009**, *31*(17), 2275−2314, 2297.
7. Biggs, J. *Teaching for Quality Learning at University*, 2nd ed.; Society for Research into Higher Education & Open University Press: Buckingham, U.K. and Philadelphia, PA, U.S.A., 2003; Chapter 2, pp 11−33.
8. Meyers, N. M.; Nulty, D. D. *Assess. Eval. Higher Educ.* **2009**, *34* (5), 565–577.

9. SCALE-UP. http://scaleup.ncsu.edu (accessed July 2010).
10. Dale, E. *Audio-Visual Methods in Teaching*, 3rd ed.; Dryden Press: New York, 1969.
11. Lord, T. *J. Coll. Sci. Teach.* **2007**, *37* (2), 14–17.
12. *Chemistry: A Molecular Science & Chemistry: A Quantitative Science*, 3rd ed.; Wertz, D., Ed.; Pearson Prentice Hall: New Jersey, 2008.
13. Oliver-Hoyo, M. T.; Allen, D. *J. Chem. Educ.* **2002**, *79* (4), 459–461.
14. Oliver-Hoyo, M. T.; Allen, D. *J. Chem. Educ.* **2004**, *81* (3), 441–448.
15. Rennie, L. J.; Punch, K. F. *J. Res. Sci. Teach.* **1991**, *28*, 193–209.
16. Laforgia, J. *Sci. Educ.* **1988**, *72*, 407–421.
17. Oliver-Hoyo, M. T.; Allen, D. *J. Chem. Educ.* **2005**, *82* (6), 944–949.
18. Eddy, R. M. *J. Chem. Educ.* **2000**, *77*, 514–517.
19. Burr, J. A.; Nesselroade, J. R. In *Statistical Methods in Longitudinal Research: Principles of Structuring Change*; von Eye, A., Ed.; Academic Press: Orlando, FL, 1990; pp 3–34.
20. Iqbal, H. M.; Shayer, M. *J. Res. Sci. Teach.* **2000**, *37*, 259–274.
21. Oliver-Hoyo, M. T.; Allen, D. *J. Coll. Sci. Teach.* **2006**, *1*, 42–47.
22. Oliver-Hoyo, M.; Justice, J. *J. Coll. Sci. Teach.* **2008**, *3*, 62–67.
23. Bloom, B. S. *Taxonomy of Educational Objectives: The Classification of Educational Goals. Handbook I: Cognitive Domain*; David McKay Company, Inc.: New York, 1956.

Chapter 5

Development and Refinement of a Research Study Assessing Student Attention in General Chemistry

Kelly Y. Neiles,[1] Elizabeth A. Flens,[1] Diane M. Bunce,[*,1] and Michael Ferguson[2]

[1]Chemistry Department, The Catholic University of America, Washington, DC 20064
[2]Science Department, DeMatha Catholic High School, Hyattsville, Maryland 20781
[*]E-mail: bunce@cua.edu

Planning an experiment and collecting and analyzing data is only half the process of doing research. Progression from research ideas to a published manuscript is an iterative process. The purpose of this chapter is to describe the decisions in that process including the development and refinement of the research questions, methodology, data analysis, and conclusions that must be completed before an article is likely to be accepted for publication. These decisions will be discussed within the context of a research experiment dealing with an innovative use of a measurement tool to collect data on student attention lapses in general chemistry. Use of reviewer comments in the journal submission process to improve the manuscript is discussed in terms of the changes that were made in the final version of the manuscript prior to acceptance for publication.

Introduction

Journal articles often sound as if the researchers progressed in a straight path from the research idea to conclusions with everything else impeccably planned and executed. This can be daunting to a new researcher. The purpose of this chapter is to provide the reader with a more realistic view of how one of our published manuscripts progressed from initial idea to published article. Inherent in this iterative process is a series of decisions that we made as researchers, sometimes in response to theory and other times in response to the reality of the research situation. When we thought we had a well written and polished product, we submitted it to a journal for review. It was surprising to us that the reviewers had so many suggestions for improvement on what we had considered to be a finished product. After we reflected on the reviews, we too agreed that we could improve our product.

The following section discusses how we transitioned from an initial idea to a research question based upon the literature on this topic.

Theory and Literature Review

As teachers, we have probably all used lectures as part of our teaching approach. It is curious that student achievement does not always reflect the quality of those lectures. The question arises as to whether or not the students were paying attention in lecture. Lapses in student attention have been thought to be a result of passive lectures where students are not actively involved in the learning process. Books on how to lecture effectively suggest that if lectures are passive presentations of information then student lapses are more likely to occur (*1*). The possible reasons for these lapses in student attention may be a result of how the brain operates and how learning takes place. For instance, student attention lapses may be a function of an overload of the working memory; student lack of interest in the topic; or the passive role of the student in the teaching pedagogy used.

Limited working memory, one of the possible reasons for student attention lapses, is the term used to describe the process by which incoming information is integrated with information retrieved from the student's long term memory. Too many stimuli presented at once without allowing time for students to process the incoming information can overload the working memory capacity (*2*). The working memory has been shown to have a capacity of 7±2 items. This capacity can quickly be exceeded in a 50 minute lecture through the normal process of listening, taking notes and thinking about multiple concepts without time to process and clear the working memory space (*3*). If working memory capacity is exceeded, then students will have a great deal of difficulty paying attention. Fluctuations in students' attention may reflect the students' efforts to process the continuing flow of information while dealing with the limitations of working memory.

Another reason for students' inability to maintain attention in lecture may result from students' level of interest in the topic. Here the teacher's role is crucial. The teacher may be able to influence students' initial interest level in a topic by helping them see its relevance. Interest according to Ainley, Hidi, & Berndorff (*4*) is characterized by focused attention, increased cognitive and affective functioning, and persistent effort. As a result, if students are interested in a topic, this may result in increased attention during lecture.

Even if students are interested in a topic, the length of a typical lecture can be overwhelming. Students in a passive mode can experience attention lapses even if they are interested in the topic. Brophy (*5*) attributes this lack of attention to students not being required to use their cognitive abilities during lecture. Others agree that attention capacity can vary depending on students' motivation to learn the information presented (*1*). A passive learning experience allows students to become uninvolved. Active learning experiences draw students into the learning process and encourage them to learn the information (*6*). These learning experiences prevent students from becoming passive and thus may be responsible for helping engage students in learning concepts that they may not have been initially motivated to learn. In essence, the teacher in an active learning process has involved students in a manner that a passive lecture may not (*6*).

Other variables affecting student attention can be more personal in nature and may include lack of sleep, poor nutrition, general health, or complications from personal relationships, among others. These variables are more individualistic in nature and harder to measure.

Previously attention lapses were identified by researchers who either observed or videotaped students during lecture (*7*). Here the occurrence of attention lapses were determined by observing a student's facial expressions. A possible threat to the validity of this approach is the fact that the student's facial expression may not be a true reflection of an attention lapse. Student input in the determination of attention lapses is crucial. Measuring student attention lapses must include accurate identification of lapses. In addition the measurement technique should include unobtrusive methods of data collection and data measured over an extended period of time, occasions, and with different teaching pedagogies. Such parameters call for new methods of data collection using more automated systems so that multiple data points can be recorded with minimal interference to the learning process.

Development of Questions and Methodology

As teachers we sometimes notice that students do not appear to pay attention during the entire length of a class. Even as faculty members in a seminar audience we may find it difficult to pay attention during the entire seminar. Are such lapses part of human nature? If so, then how often do such attention lapses occur? We were interested in investigating these questions in regards to students enrolled in general chemistry classes. We were also interested in investigating whether the number of student attention lapses is affected by using different teaching pedagogies within a given class.

A search of the literature produced few studies where student attention lapses were actually measured but quite a few references where definitive advice was offered on how to increase student attention during lecture (*1*, *8*). This situation increased our interest in developing both a way to measure student attention lapses during a class and the effects of different teaching methods on these lapses. Once we decided the main focus of our research, we developed the following four main research questions (*9*):

1. Does student attention remain constant during a general chemistry class?
2. Are there differences in the length of attention lapses reported?

3. Is there a difference in student attention during different teaching pedagogies within a given class?
4. Is there a difference in student attention during the beginning, middle or end of a class?

Although we were satisfied with our progress in developing the research questions to this point, it soon became obvious that a number of decisions regarding the research project were in order. These decisions included developing operational definitions of parameters involved in this study. To develop valid terminology that students could relate to, we invited general chemistry students to meet with us and discuss their understanding of the descriptions we would later use to describe our study to the students in the study. As a result of this meeting, we learned that the term "attention lapse" was not as effective in communicating our plan to students as was the phrase "zoning out". The students we interviewed also discussed what might constitute short, medium or long lapses of attention in a class. This discussion helped insure that the student directions we wrote for the study would be better understood by the student participants.

We realized too, that we needed to differentiate among commonly used terms such as "lecture", "lecture segment", "class", and "course" for this study. Although these terms are sometimes used interchangeably in casual conversation, in our experiment, they had specific and unique meanings. We recognized the need to be consistent in our use of these operationally defined terms. Our operational definitions for these terms are presented here (*9*):

Lecture: traditional pedagogical approach involving the teacher presenting information to an audience. The flow of information proceeds from teacher to student.

Lecture segment: portion of the class devoted to lecture pedagogy.

Demonstration: use of chemicals or models to present a visual presentation of the chemistry concept being presented.

Clicker Question: ConcepTests (*10*) presented electronically through the use of personal response devices.

Class: the full length teaching session. In this study, all classes had a duration of 50 minutes.

Course: the semester-long curricula, which for this research included general chemistry for engineering students; general, organic and biochemistry for nursing students; and chemistry for nonscience majors.

Pedagogical approach: any unique presentation or interaction between teacher and student during a class. Typical examples include lecture, demonstration, clicker questions, working in student groups/pairs, inclusion of real world applications, personal vignettes, or announcements.

Sample

To increase the generalizability of the study, we decided to sample a range of general chemistry courses including general chemistry courses for engineering, nursing and nonscience majors. In our institution, these three populations have separate general chemistry courses which we will refer to as Chem I (engineering students) (n= 74); Chem II (nursing students) (n= 68); and Chem III (nonscience majors) (n= 44). Including all three of these courses in our research had the added advantage of using courses taught by more than one teacher. Chem I was taught by teacher A and teacher B taught both Chem II and Chem III. The research idea was presented to the two teachers and their cooperation was secured.

Institutional Review Board (IRB)

All research involving human participation is required to be reviewed and approved by an Institutional Review Board (IRB). Research is judged as being either exempt or nonexempt from the full IRB committee review according to the guidelines set by the National Institutes of Health. Our research was determined to be exempt because only normal educational practices were employed in the study and no student was identified through the collection and reporting of data. Another key element of our exempt status was that all participation was voluntary and participant identity was not revealed to the teachers in the study.

Methodology and Data Collection

Personal Response Devices (Clickers)

In an effort to measure self-reported student attention lapses during class over an extended period of time, a data collection process was devised that would cause minimal disruption in each of the courses. Our familiarity with personal response devices (clickers), led us to the idea that students' reports of attention lapses could be collected through the use of a dedicated class set of clickers

A research proposal was submitted to the clicker company who agreed to a no cost one-semester loan of a class set of clickers plus a radio frequency receiver for use in this study. Responses from these clickers were recorded by a receiver on a laptop tablet PC in the back of the room. This tablet PC was also used by the researchers to record different pedagogies used by the teacher during each class. Student clicker responses were surveyed every 30 seconds by setting up a PowerPoint presentation on the tablet PC that automatically advanced a PowerPoint slide every 30 seconds. Changes in teaching pedagogy were recorded with the tablet PC stylus on the appropriate slide in that PowerPoint file.

In two of the courses studied, students owned a personal clicker and used it regularly in class. In the third course, students did not own personal clickers. All three courses were provided with a set of clickers on lanyards for this research study. Students used the lanyard clickers exclusively to record attention lapses. This class set of lanyard clickers was distributed at the beginning and collected at the end of each class.

Directions to Students

Students were instructed to select Button #1 on the lanyard clicker if they believed their attention lapse was 1 minute or less; Button #2 if the lapse was 2 to 3 minutes and Button #3 if the lapse was 5 minutes or more. To help students understand what typical lapses of these durations might be, examples of each type of attention lapse were suggested. These suggested attention lapses were based on examples that the students we consulted at the beginning of the project provided. For example, Button #1 (lapse of 1 minute or less) might include looking at a clock/watch, reading a text message, or daydreaming. Button #2 (2 to 3 minute lapse) might be typing a response to a text message. Button #3 (5 minutes or more) might include working on assignments for other classes or falling asleep.

Students were reminded in class on a regular basis to use their lanyard clickers to record attention lapses when they occurred. Reminders were delivered verbally when clickers were handed out at the beginning of class, by the instructor during class, or as a footnote on the teacher's PowerPoint slides used in class.

Identifying Teaching Pedagogies

Inter-rater reliability was calculated on the start time and duration of the different pedagogies used during the class. The use of several identifiable pedagogies within a single class period resulted in a class being divided into pedagogical segments. For example, a single class could contain several segments of different pedagogies including multiple segments of a specific pedagogy. During the first week of data collection, inter-rater reliability did not meet acceptable research standards. Meetings were held with the researchers to review what constituted a pedagogical change. This resulted in an acceptable inter-rater reliability statistic during the remaining data collection.

Other Variables

In two of the three courses (Chem II (nursing) and Chem III (nonscience)), data on additional variables were collected. In these courses, students routinely completed a diagnostic test of logical thinking called the Group Assessment of Logical Thinking (GALT) test during the first two weeks of the course (*11*). The GALT is a 12-question Piagetian test that measures students' use of logical thinking. The GALT test was delivered electronically in this experiment. The possible scores range from 0 to 12, and for 10 of the 12 questions include the selection of a correct answer and a correct reason for that answer. The last two questions involve the ability to group variables. Student GALT scores were graphed and natural breaks in the score distribution were used to establish low (1-6), medium (7-9), and high (10-12) GALT scores.

In these same two courses (Chem II (nursing) and Chem III (nonscience)), final course averages were used as indicators of achievement. Student averages were graphed and natural breaks in the distribution were used to establish low (0-75%), medium (76%-85%), and high (86%-100%) achievement scores. Gender was also noted in these two courses.

GALT, gender and achievement could not be used for the third course (Chem I (engineering)) because clicker use in this course was not part of the regular teaching scheme and no clickers were registered to any particular student. Thus, identifying data could not be recorded and specific students could not be tracked.

Timeline

Data were collected in each course three days a week for 6 weeks. Data from the first two weeks of the experiment were used in a formative manner to both help refine the methodology and familiarize students with the use of clickers. The remaining 4 weeks of data collection were used in the analysis.

Data Analysis

Data Reduction

Data collected on some days were eliminated from the analysis due to incomplete data sets. This was a result of the teacher scheduling short lab experiences during class, use of group worksheets, or researcher error in data collection on that day. The long data collection time period of this research helped ensure that such glitches did not seriously affect the integrity of the data.

This particular experimental design provided a relatively large data set. Although this was a plus, it also meant that we had a large, almost overwhelming data set to analyze. As in most experiments, a method had to be devised to reduce the data in a way that was in accord with the prerequisites of the statistical methods used. Data reduction involved refinement of research questions and the definition of conditions under which these questions were addressed.

To start the data reduction process, we reviewed the number and frequency of pedagogies used in the three courses of this study. We documented fourteen types of pedagogies and made the decision to analyze only the three most frequently used pedagogies. These three pedagogies common to at least two of the three courses were lecture, clicker questions, and demonstrations. Since students recorded more than one response per class over multiple days, the appropriate statistic to analyze these data is a repeated-measures ANOVA (*12*). This statistical procedure is predicated on the analysis of equal segment lengths. The longest common segment of lecture, clicker question session, or demonstration that was common to all classes within a course was chosen as the unit of analysis. When different pedagogies were compared to lecture segments, a new common segment length for both the pedagogy and lecture segment was established as the unit for that analysis. An in-depth discussion of the analyses used in this study can be found in Bunce, et al (*9*).

Refining the Research Questions

Once we looked at the data initially, we developed additional questions we could ask of the data. This led us to a series of subquestions that could be addressed through the statistical procedures we would use. The additional questions that we asked of the data in relation to the four general questions are in bold face and included in the following (*9*):

1. Does student attention remain constant during a general chemistry class?
   a. **Is this attention different for different gender, GALT or achievement levels?**

2. Are there differences in the length of attention declines reported?

3. Is there a difference in student attention during different teaching pedagogies within a class?
   a. **Does the use of different pedagogies affect the attention lapses reported during subsequent pedagogies?**

4. Is there a difference in student attention during the beginning, middle or end of a class?
   a. **Is there a difference in the number of short, medium and long duration attention lapses during short, medium, and long lecture segments?**

Checking Assumptions of Statistical Procedures

If statistical procedures are used without first checking that the data do not violate the assumptions of the statistical procedure, misleading interpretations of the statistic can result. Violation of assumptions for a statistical procedure can range from a minimal effect on interpretations of results to the fact that the data may not be appropriate for a given statistic. Good practice requires that researchers be aware of violations of assumptions if present and either choose a different statistic or modify their conclusions accordingly.

In this experiment, the choice of the repeated measures ANOVA required checking the following data assumptions: 1) independence of observations; 2) normal distribution of data; and 3) homogeneity of variance (*12*). Our data set met all three assumptions as described in Bunce, et al (*9*).

Results and Conclusions

The results presented here are a summary of those reported in Bunce et al. (*9*). Repeated measures ANOVAs were used to address the questions we investigated as follows (*9*):

1. Does student attention remain constant during a general chemistry class?

The three repeated measures ANOVAs (one for each course, Chem I (engineering), Chem II (nursing), and Chem III (nonscience)) produced the following results presented in Table I:

Table I. Student Reported Attention Lapses During Class

Course	*F (df)*[a]	*Significance*
Chem I (engineering)	1.22 (26, 2863)	0.205
Chem II (nursing)	1.60 (19, 1313)	**0.047**
Chem III (nonscience)	0.86 (9, 641)	0.571

[a] Degrees of freedom reflect the number of student responses over the course of the experiment and not the number of students enrolled in the course. Bolded value is significant at $p<0.05$.

The results show that there is a significant difference in self-reported student attention lapses during a class in only one course (Chem II (nursing)). This result may be due to the fact that the largest student participation occurred in this course compared to the others. Chem II had an average daily participation of 56% compared to Chem I (engineering) of 23% and Chem III (nonscience) of 27%. This significant result in Chem II can be interpreted as showing that there is a fluctuation between attention and attention lapse during the class.

1a. Is there an interaction between student attention and each of the following: gender, GALT and achievement?

In this study, the one course (Chem II (nursing)) that demonstrated a significant effect and which had the largest student participation was further analyzed for interaction effects with gender, GALT and achievement. The results indicate that there is no significant difference between male and female reports of attention lapses in Chem II, $F(1,49)=0.01$, $p=0.917$. When GALT is used as a control variable, there is no significant difference in the reporting of attention lapses among low, medium, and high GALT groups for Chem II, $F(2,49) = 1.02$, $p=0.369$. When achievement is entered as a control variable, no significant difference in self-reported attention lapses was found among low, medium, and high achieving groups for Chem II, $F(2,49)=2.21$, $p=0.121$. These results suggest that attention lapses are independent of gender, GALT level, or achievement for this course.

2. Are there differences in the length of attention declines reported?

Since Chem II (nursing) was the only course that showed significant differences in the overall reporting of attention lapses throughout lecture segments, only data from this course was used to investigate the question regarding length of lapses. Students registered differing lengths of attention lapses by pressing one of three buttons on their research lanyard clickers. A repeated-measures ANOVA was used to compare the means of these responses of different attention lapse length. The results indicate that the shortest attention lapse (1 minute or less) was reported by students at a significantly higher rate than the medium (2-3 minute) and long lengths (5 minutes or longer), ($F(4,324)=8.30$, $p=0.000$). This suggests that the most prevalently reported attention lapse is of short duration (1 minute or less).

3. Is there a difference in student attention during different teaching pedagogies within a class?

To analyze this question using a repeated measures ANOVA, students' reporting of attention lapses during student-centered pedagogies (demonstrations and clicker questions) were compared to their reporting of attention lapses during lecture segments within the same class. The pedagogies used in each course are presented in Table II. In one of the courses, Chem III (nonscience), all three pedagogies (lecture, demonstration and clicker questions) were used by the teacher. In Chem I (engineering), the teacher used only lectures and demonstrations, and in Chem II (nursing), the teacher used lectures and clicker questions. This resulted in analyses of only one student-centered pedagogy (demonstration or clicker question) for Chem I (engineering) and Chem II (nursing) compared to lecture segments. In Chem III (nonscience) both demonstrations and clicker question pedagogies were compared to lecture segments.

Table II. Pedagogies Used in Each Course[a]

	Lecture	*Clicker Questions*	*Demonstrations*
Chem I	x		x
Chem II	x	x	
Chem III	x	x	x

[a] Reprinted with permission from Bunce, D. M.; Flens, E. A.; Neiles, K. Y. How long can students pay attention in class? A study of student attention decline using clickers. *Journal of Chemical Education*, **2010**, *87* (12), 1438-1443. Copyright 2011, American Chemical Society.

In Chem II (nursing) and Chem III (nonscience), the effect of the clicker pedagogy was significant (Chem II (nursing), $F(1,67)=26.71$, $p=0.000$, and Chem III (nonscience), $F(1,67)=6.93$, $p=0.011$). In Chem I (engineering) and Chem III (nonscience), the effect of the demonstration pedagogy was significant (Chem I (engineering), $F(1,86)=7.22$, $p=0.009$ and Chem III (nonscience), $F(1,67)=5.46$, $p=0.022$). These results indicate that the use of student-centered pedagogies (demonstration or clicker questions) significantly reduces the number of attention lapses reported by students when compared to the number of lapses reported during lecture segments.

3a. Does the use of student-centered pedagogies affect attention lapses during subsequent lecture segments?

To address this question, only portions of lecture segments both before and after student-centered pedagogies (demonstrations or clicker questions) were used in the analysis. Lecture segments had to be equivalent in length to that of the student-centered pedagogy to satisfy the prerequisites of the statistic. A repeated measures ANOVA was used in this analysis. In Chem II (nursing) where clicker questions were used, a significant decrease in the number of attention lapses was reported in lecture segments that occurred *after* versus *before* the use of a clicker question ($F(1,66)=8.70$, $p=0.004$). In Chem III (nonscience) where demonstrations were used, a significant decrease in the number of reported attention lapses was also found in lecture segments that occurred *after* versus *before* a demonstration ($F(1,64)=4.25$, $p=0.043$). These results indicate that the use of student-centered pedagogies decreases the number of reported attention lapses not only during the pedagogy itself, but also in lecture segments that immediately follow the student-centered pedagogy compared to those in the lecture segments prior to the examined pedagogy.

4. Is there a difference in student attention within a lecture segment at the beginning, middle or end of a class?

In Chem II (nursing), a repeated measures ANOVA showed an overall significant difference in the reporting of attention lapses in the lecture segments at the beginning, middle and end of class ($F(2,324)=12.78$, $p=0.000$). A post hoc analysis indicated a significant difference between the beginning and middle of a class with a significantly higher number of attention lapses reported during the middle of a class. A similar effect is seen between attention lapses reported in the lecture segments in the middle vs. the end of a class though it is not a significant difference. This is interpreted as the middle part of a class having a significantly higher number of self-reported attention lapses than a lecture segment at the beginning of a class.

4a. Is there a difference in the length of self-reported attention lapses during lectures of differing lengths?

In Chem II (nursing), the length of student reported attention lapses was compared within lectures of differing length. This analysis investigated whether the length of attention lapse was affected by the length of the lecture segment. The results of this analysis are presented in Table III. Overall, there is a significant difference in the length of attention lapses reported in lecture segments of varying lengths ($F(4,324)=8.30$, $p=0.000$). Due to the small n and the large number of possible comparisons which would result in a high Type I error, post hoc analysis was not pursued. Based on the means of attention lapses in lectures of varying lengths, the trend appears to show that short attention lapses are the most prevalent type of attention lapse in lectures of different lengths. Together with the previous analysis that showed a significantly higher occurrence of short attention lapses throughout the length of an entire class, this result showing a trend for short attention lapses during lecture segments of varying lengths appears valid.

Table III. Difference in the Length of Attention Decline in Short, Medium, and Long Lectures in Chem II (Nursing)

Length of lecture	*Length of attention decline*[a]	*Mean*	*F (4,324)*	*Sig.*
Short	1	1.62	8.30	**0.000**
	2	0.41		
	3	0.09		
Medium	1	5.18		
	2	1.65		
	3	0.63		
Long	1	3.23		
	2	1.12		
	3	0.59		

[a] Length of attention decline: 1=1 minute or less; 2=2-3 minutes; 3=5 minutes or more
Bolded value is significant at $p<0.05$.

Discussion of Results

In general, the analysis of the data was limited to one course (Chem II nursing) that showed a significant overall effect for attention lapses over time (*9*). This was also the course with the highest daily average participation. Similar but nonsignificant trends were also evident in the two other courses Chem I (engineering) and Chem III (nonscience).

Limiting the discussion to Chem II (nursing), we see that student attention fluctuates over time during the length of a typical class. This fluctuation peaks with the largest number of self-reported attention lapses typically occurring during the middle of the class. The peak during the middle of class is significantly higher than self-reported attention lapses at the beginning. Based on this result, teachers should be aware that student attention lapses are most likely to occur during the middle of a class and plan to re-engage students at this point.

Within a lecture segment, the majority of self-reported attention lapses were characterized by students as short duration (1 minute or less) regardless of the length of the lecture segment studied. This means that when students experience attention lapses, these lapses are likely to be short and students should be able to reestablish their attention quickly. Since students operate individually in terms of their pattern of attention and attention lapses, a teacher is faced with some students being engaged at a given moment and others who are not. Teachers may not be aware of the attention and attention lapse cycle of the students in the class because they may be focusing on only those students who appear to be fully engaged or non-engaged in the lecture at any one time.

When gender, GALT (logical reasoning ability) level and achievement variables are investigated for two courses (Chem II (nursing) and Chem III (nonscience)), none of these three variables significantly affects the reported attention lapses. For example, women do not report more or fewer attention lapses than men at any given time within a lecture segment. Students who perform at a low GALT (logical reasoning ability) level do not report a significantly different number of attention lapses than students who perform at a high or middle GALT level. This is also true for students of differing achievement levels. All students regardless of gender, logical reasoning ability, and achievement report comparable number of attention lapses during lecture segments.

In the analysis concerning the use of different teaching pedagogies (demonstrations and clicker questions), data from all three courses were used. Here, self-reported attention lapses for the two student-centered pedagogies compared to lecture segments of comparable length, are significantly fewer than during lecture segments. This significant reduction in the number of attention lapses is still evident in lecture segments that follow the use of these pedagogies. This suggests that these alternatives to lecture (demonstration and clicker questions) are more successful at engaging students and reducing the occurrence of attention lapses. This positive effect carries over to subsequent lecture segments. Teachers who use a variety of student centered pedagogies during a given class could expect increased student attention overall.

The Review Process

No matter how perfect a manuscript appears to the authors when submitted for publication, objective reviewers are often able to point out inconsistencies or deficiencies in the manuscript. Authors very often are resistant to these critiques but if a manuscript is not able to convey a complete and convincing scenario to someone not directly involved in the research, then there is room for improvement. Knowing how to interpret and respond to such reviews is an important learning process for the researcher.

In this manuscript, the description of the analysis and results was complex. The reviewers asked pointed questions that helped us understand where we were losing the reader. We were able to develop a more expanded description of what we did statistically and explain it so that someone not familiar with the statistics used could better understand the analysis. The data reduction process in the originally submitted manuscript was one area that also needed more explanation. In the methodology section, we had to more fully explain our method of data collection that used research clickers and a receiver to collect data automatically every 30 seconds within a PowerPoint framework. Questions concerning the reported percentage of student responses also required more clarification. The usual formatting and overlooked grammatical errors were included in the reviews. Although seemingly painful to construct, the resulting revised manuscript was considered an improvement by the authors as well as the reviewers.

Published reports of research may present a polished, coherent picture executed through a series of seemingly sequential steps. The reality of doing research from conception, data collection, analysis, submission, revision and ultimately to publication do not always follow such a direct path. New researchers may feel that they are not successful when their research does not seem to follow a direct path. In reality, very few people's research does. The secret to becoming a successful researcher is to keep trying and learn something new with each turn in the road to completing and publishing research.

Acknowledgments

We would like to thank the teachers and students who participated in this study and Turning Technologies, Inc. for the loan of the class set of clickers used in this study.

References

1. McKeachie, W. J. *McKeachie's Teaching Tip*; Houghton Mifflin Company: New York, 2002.
2. Miller, G. A. *Psych. Rev.* **1956**, *63*, 81.
3. Dehn, M. J. *Working Memory and Academic Learning: Assessment and Intervention*; John Wiley & Sons, Inc: Hoboken, NJ, 2008.
4. Ainley, M.; Hidi, S.; Berndorff, D. *J. Ed. Psychol.* **2002**, *94*, 545–561.
5. Brophy, J. *Motivating Students to Learn*, 2nd ed.; Lawrence Erlbaum Associates: Mahwah, NJ, 2004.
6. McCombs, B. L.; Miller, L. *Learner-Centered Classroom Practices and Assessments*; Corwin Press: Thousand Oaks, CA, 2007.
7. Johnstone, A.; Percival, F. *Educ. Chem.* **1976**, *13*, 49–50.
8. Bligh, D. A. *What's the Use of Lectures?*; Jossey-Bass: San Fransisco, CA, 2000.
9. Bunce, D. M.; Flens, E. A.; Neiles, K. Y. *J. Chem. Educ.* **2010**, *87*, 1438–1443.
10. Mazur, E.; Somers, M. *Am. J. Phys.* **1999**, *67*, 359–360.
11. Roadrangka, V. The Construction and Validation of the Group Assessment of Logical Thinking (GALT). EdD. Dissertation, University of Georgia, 1985.
12. Weinberg, S. L.; Abramowitz, S. K. *Statistics Using SPSS: An Integrative Approach*, 2nd ed.; Cambridge University Press: New York, 2008.

Chapter 6

Teaching Chemistry with Visualizations: What's the Research Evidence?

Vickie M. Williamson*

Texas A & M University, College Station, TX 77843-3255
***E-mail: williamson@tamu.edu**

The use of visualization techniques to teach chemistry has been touted in the literature, with the idea that these techniques promote the formation of mental images in students, which depict particle behavior during chemical processes and help better connect the macroscopic and particulate natures of chemical phenomena. However, a number of myths or misconceptions concerning the use of visualizations in the classroom exist. These myths or misconceptions can prevent instructors from using visualization techniques. This chapter seeks to provide evidence concerning these myths, so that instructors can knowledgeably plan to integrate visualizations into their courses.

Background

A visualization can be defined as: "1. the creation of a clear picture of something in the mind, 2. a clear picture of something created in the mind, or 3. a technique whereby somebody creates a vivid positive mental picture of something such as a desired outcome to a problem" (*1*). In chemistry, mental pictures or visualizations of particle behavior are required for students to understand particle behavior. Understanding particle behavior is one of the three basic components of chemistry, as described by Johnstone (*2*). Johnstone's components of chemistry include the macroscopic representation (observations that can be made with the human eye, such as in the laboratory or everyday life), the symbolic representation (chemical and mathematical symbols to represent the phenomena), and the submicroscopic representation (particles or atoms and molecules) (*2*).

It is expected that chemists or experts can effortlessly move among these three representations of a chemical phenomenon in their thinking.

Some researchers have found that student misconceptions of chemistry can be attributed to students' inability to visualize particle behavior (e.g., (*3–5*)). These misconception findings were followed by studies that investigated various techniques to help students visualize particles. These techniques will be described in the following section. So, what can we do as the instructor to encourage our students to visualize particle behavior?

What type of visualization will work in the classroom?

Some instructors are not aware of the wealth of literature concerning various visualizations techniques. In their synopsis of the literature, Williamson and Jose (*6*) outlined six general techniques to promote visualization of particles in students' mental models. These included the use of physical models, role-playing, fixed computer models, dynamic computer models or animations, student-generated drawings or animations, and interactive computer models/animations. Each will be described below, along with literature examples.

Benefits from the use of physical models have been long accepted in chemistry. Talley (*7*) found that freshman college students who used ball-and-stick models throughout the semester were better able to visualize chemical concepts, transfer particulate understanding, and perform better on general chemistry assessments requiring critical thinking. Talley advocated using models throughout the semester to show both structure and interactions. While instructor-built models may be fine to start with, the consensus is that students need to build their own models (e.g., (*7–10*)). Wu and Shah (*8*) concluded in their review of the literature that students who watched teacher demonstrations with concrete models were able to better visualize molecules, but that it was crucial for students to manipulate the models themselves in order to better solve chemistry problems and better represent chemical concepts as particles or symbols. Nicoll (*9*) reported success using Play-Doh to model molecules and proposed that traditional model kits may lead students to correct geometries, but limit other aspects that students may show like bond length or lone pairs. Other things like gumdrops, raisins, and marshmallows with toothpicks or magnets on a metal surface have also been used. No matter the medium used, students should be allowed to rotate the models in order to see the angles and any changes in the model during a chemical process thus promoting connections between the model and chemical processes (*7–10*).

Role-playing is an activity where the students are asked to portray atoms or molecules. While some evidence exists that role-playing is helpful to understanding, it has not been as thoroughly investigated; perhaps because it is not a technique that some instructors feel match their teaching style. Battino (*11*) describes role-playing in his college class as a participatory demonstration, which leads students to quickly understand abstract concepts and actually saves lecture time. Lerman (*12*) describes increased student interest and retention through the use of projects that allow students to role-play chemical processes. Concerns include the facility space required and whether students will participate. Battino (*11*) cited a minimum of 10 x 20 ft space needed for a large lecture class and that positive reinforcement would keep students participating throughout the course. Further, multiple uses of role-playing throughout the course are required (*11*, *12*).

Fixed computer models include those programs that allow a student to view and perhaps rotate a computer depiction of a molecule with fixed coordinates. This type of visualization technique can be achieved using a number of programs (e.g., Spartan, ChemBioDraw, or jmol). Barnea and Dori (*13*) found that high school students who used the computer models had greater improvement in their understanding of structure and bonding, along with their spatial ability than students who simply built ball-and-stick models. In an organic chemistry unit, Dori and Barak (*14*) found that using a fixed computer-modeling program was significantly better than a teacher demonstrating the ball-and-stick models. Students in the computer group were able to generate better explanations of molecular structure. The hard copy version of fixed computer models involves the use of static pictures of particles. During three two-week topics of study, Bunce and Gabel (*15*) reported that teaching with static representations of molecules, along with macroscopic and symbolic representations, resulted in significantly higher problem solving than the control group, which did not have the particulate representations. Gender effects also were seen in this large study. Females taught with the particulate representations scored statistically equilvalent to the males, while females in the control group scored significantly lower than the males. The achievement of males did not seem to be affected by teaching with static paper-and-pencil particulate representations. With fixed computer models (or static particle pictures), students are not allowed to show their own understanding of the concepts at the particulate level, as the drawings are preset whether on the paper or in the computer program. In the case of most computer programs, students can, at most, only rotate a molecule with preset angles. See Figure 1 for an example of a fixed computer model.

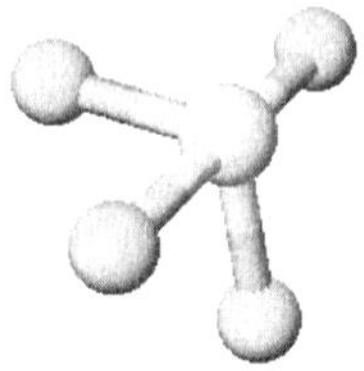

Figure 1. The SF4 molecule in J-Mol.

Dynamic particulate computer models or animations show a series of images to depict motion. These animations are movies that can be projected during lecture or viewed on the students' computers. The benefits of animations over static pictures are thought to be that animations prompt the formation of dynamic mental models (e.g., (*16*, *17*)). Animations are the most widely used of the six methods discussed here. As a result, many of the myths held by instructors deal with the use of animations. When later discussing these myths, the chapter will present research findings concerning animations.

Student-generated drawings or animations help to expose the mental models that students hold of particle behavior (*18–20*). Asking a student to create a drawing helps to elicit the student's mental model on paper of the chemical behavior, and thus can be used to drive the instruction. A number of researchers have used these drawings to gage student understanding (*18–20*). Benefits from student-generated animations have been reported from low-tech methods, such as the creation of a flipbook or a set of storyboards (*21*) to higher-tech methods, such as the creation of a computer animation (*22*). The free ChemSense program (*22*) (available at http://chemsense.org) was developed for use in accessing understanding and is easy for both instructors and students to use. See Figure 2 for an image of the ChemSense Animator. More professional software programs tend to have steep learning curves that limit their usefulness with beginning chemistry classes. A few studies have investigated student-generated drawings vs. student-generated animations. At least one study found no significant differences in understanding gains from using storyboards vs. the creation of computer animations with college chemistry students on a two-week unit on equilibrium (*23*). The concern with student-generated drawings or animations is that even with the use of a good rubric, these visualizations are difficult for the instructor to evaluate.

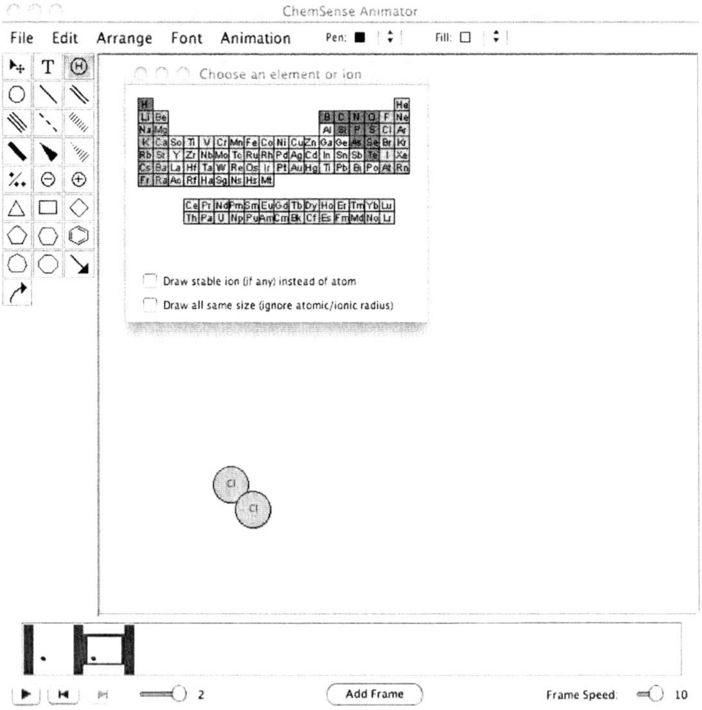

Figure 2. ChemSense Animator.

Interactive computer animations allow students to control the variables that direct the action of particles on the screen. During an NSF workshop, experts in the field called for the classroom use of regular animations to demonstrate particulate behavior. They also called for the use of interactive computer animations or models, which allows students to modify the action of particles (*24*). Laboratory simulations at the macroscopic level where students control the variables may be interactive, but are not necessarily tied to the particulate level. Some interactive models begin as a laboratory simulation, and then switch to a particulate-level view of the phenomena. This is a fairly new technique, so although there are announcements of these computer applications/programs in the literature, little research has been done to test their effectiveness. For example, Abraham, Gelder, and Haines (*25*) report that their gas law interactive simulation provides a particle-level view with a graphical display and variables which are controlled by the student and can be used to collect data for inquiry-based investigations. The authors are working on other simulations and evaluation of their inquiry activities using the simulations. See Figure 3 for an example.

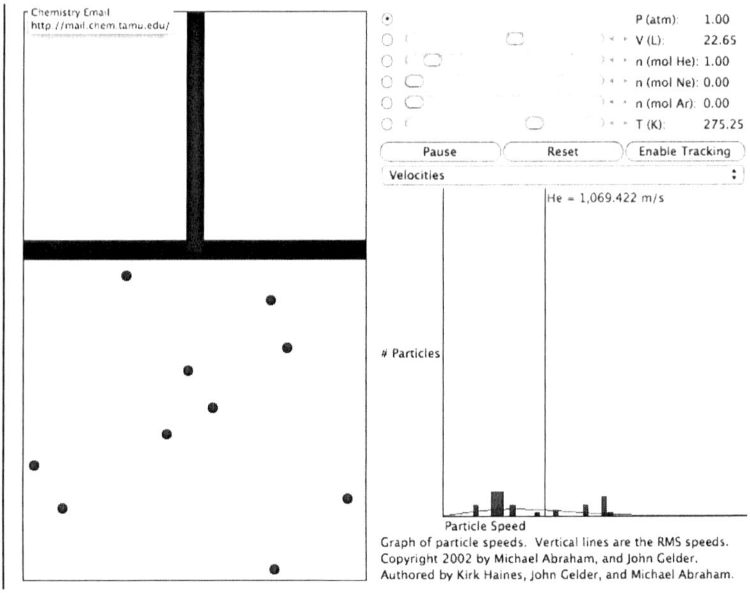

Figure 3. Interactive Computer Animation (screenshot accessed June 2011 from http://genchem1.chem.okstate.edu/CCLI/Startup.html).

In summary, physical models, role-playing, fixed computer models, dynamic computer models or animations, student-generated drawings or animations, and interactive computer models/animations are techniques that can be used to promote visualization of particles in students' mental models. In the previous discussion of these six general techniques summarized from the literature (*6*), only representative studies from the existing literature were cited. There are several literature reviews in this area that an interested reader should consider. Hoffler and Leutner (*26*) in their meta-analysis of 26 studies from many disciplines concluded that there was a medium sized advantage for instructional animations over static pictures. Considering visualization specifically in science education, a volume edited by Gilbert (*27*) and a second edited by Gilbert, Reiner, and Nakhleh (*28*) provide a good discussion of the roles of models and visualizations in instruction along with the underlying cognitive theory. In their review of the literature, Wu and Shah (*8*) focused on chemistry learning by examining three types of studies including: correlation studies of spatial ability and chemistry learning; student problems in understanding visual representations in chemistry; and visualization tools to help students with these problems. Finally, a volume edited by Gilbert and Treagust (*29*) focuses on chemical education and the issues encountered when multiple representations are used in the discipline.

Theoretically, why should these visualization techniques help understanding?

From a constructivist perspective, learning is an active process in which an individual constructs meaning from his/her experiences by linking new knowledge to existing conceptual frameworks (*30*). Johnson-Laird (*31*) proposes that our knowledge lies in our ability to construct mental models from our conceptual frameworks, as it is with these mental models that we can reason. Therefore, when students form particulate mental models, as they purportedly do with visualization techniques, the students should be able to reason about particulate behavior. This would explain the increase in conceptual understanding that many studies report when visualization techniques are used.

Myths

A number of myths or misconceptions exist concerning using visualization techniques in the classroom. These myths can prevent an instructor from using visualization techniques. This chapter will present the most common myths or objections to using visualizations, followed by research findings. While this is not an exhaustive list, hopefully, it will aid instructors.

Myth #1: Animations take too much time from lecture.

Many instructors feel that they have a hard enough time covering the required topics without the insertion of additional techniques. The objection that animations will take too much time from the lecture portion of the class is often raised. This is counter to the findings in this area.

Williamson and Abraham (*16*) used computer animations with first-semester college chemistry students during two-week courses of study. Students in the control group had the same instructor and saw static overheads with verbal descriptions of the same chemical phenomena. Students in the lecture treatment group saw animations during lecture with the instructor pointing out the important features. One to two of these short (1—2.5 minutes) silent animations were used daily. The researchers felt that the lecture treatment might not contain enough treatment time to be able to detect any effects, so a second treatment group was added. Students in the computer laboratory treatment group saw the animations during lecture with the lecture treatment group, but also saw the same animations during a weekly one-hour computer laboratory session. During the computer laboratory session, students could repeatedly play the animations and were required to complete a worksheet containing questions about the animations (e.g., How many molecules are on the screen?) and the chemistry phenomena (e.g., What happens to the volume of the gas as temperature increases?).

This process was completed for one two-week unit of study (Gases, Liquids, and Solids), and then was repeated after a two-week interval for a second two-week unit (Reaction Chemistry). This second unit was used because the researchers felt that the animations shown in the course might be such a novelty that students would pay more attention strictly due to the novelty. It was possible that results from the first unit would not be repeated in subsequent units once the novelty was over (the Hawthorne Effect). Results, however, were the same for both units of study. Student attendance was also recorded, as subjects used in the study could only miss one lecture in the two-week period.

The researchers found that there were no differences between the lecture treatment and the computer laboratory treatment on any measure. There was also no difference in any of the three groups on their mathematical, algorithmic examination items or on any attitudinal measure. The researchers found that the groups had no significant differences in their logical reasoning ability (proportions, combinations, correlations, control of variables, and probability) as measured by the Test of Logical Thinking (TOLT) (*32*). In the original article (*32*), a TOLT mean of 4.4 out of 10 was reported for 247 college science students. The TOLT in the Williamson & Abraham study (*16*) was higher and explained 19-27 percent of a conceptual understanding score, indicating that the effects of the TOLT score should be removed (co-varied) from the conceptual score to properly evaluate the treatment. Both treatment groups significantly outperformed the control group on a conceptual test of particle behavior (PNMET) even though the test items did NOT include the exact chemical systems as those in the animations The PNMET contained both new items and items drawn from other researchers (*4*, *33*, *34*). For this conceptual test, students in either treatment group scored about 1/2 a standard deviation higher that the control group (effect sizes of 0.53-0.59 were found). Researchers attributed the results to the consistent use of the short animations, the topics used in the study, and the higher TOLT scores of the students in this study compared to those in the original Tobin and Capie (*32*) article. The findings seem to indicate that particulate animations produced better transfer of particulate behavior understanding, probably due to the formation of dynamic, more expert-like mental models of the particles rather than the static images generated from overheads. Thus a lot of benefits resulted from a few minutes of class time.

Others have found similar results. For example: Yezierski and Birk (*35*) used four animations of water, each without a soundtrack, with students from eighth-grade to college level. The control group completed and discussed a worksheet with questions about the particulate nature of matter, while the treatment group spent the same time (25 minutes) viewing and discussing four short animations, the last of which was about 32 seconds long. Three weeks before the study, a conceptual pre-test was given. The same instrument was used as a post-test about a week after the intervention. Results showed significant improvement in the conceptual test for the animation group. Additionally, the pre-test conceptual test showed a difference in gender, with a preference for males, which was not evident in the post-test. This treatment closed the gender gap for their subjects. The authors cautioned that the treatment in this study involved more than just showing the animations; students were given time to discuss, question, and interpret the animations.

Likewise, Sanger, et al. (*36*) found in their study of 210 second-semester general chemistry students that viewing a 5-10 minute animation significantly improved their score on the particulate post-test question from that of their pre-test. Students also were asked whether any of the particles in the animation were moving differently than they expected and to explain the response they gave on the post-test. Viewing the animation resulted in significantly more correct answers on the post-test than the static drawings in the pre-test. Additionally, 94% of the students who answered the question correctly on the pre-test did not change their response to an incorrect option after viewing the animation.

Thus, for the instructor worried about the possible loss of instructional time, the previously discussed literature shows that short duration, particulate animations are effective. This means that only a few minutes of the class period are needed to add animations into the normal instructional sequence. Burke, Greenbowe, and Windschitl (*37*) list short time duration as the first in their list of characteristics for effective instructional animation sequences. Thus, instructors need not worry about the use of visualization techniques inherently resulting in an appreciable loss of instruction time. However, the studies described here give rise to another myth dealing with how animations should be used in class.

Myth #2: You can simply show animated sequences intermittently throughout the course.

This myth involves how to incorporate animations into instruction and how often to use them. Some instructors do use animations in class, but they simply play them for students with no other reference to the animation or its content. Others assign students to watch animations as homework, but never refer to them again. In some classrooms, particle-level animations are only used once or twice during the term.

Animations, like any other technique, are not used best in isolation. The role of an instructor is to help students make connections between the content in the course. The content in animations is no exception. Some researchers advocate that animations must be used in conjunction with:

- discussion to allow students to interpret the animations and to relate the animations to macroscopic chemistry behavior (*19, 35, 38*)
- active learning strategies (prediction, decision making, assessment with feedback, inquiry, etc.) (*37, 39, 40*)
- descriptions of expectations or rationale for responses (*36*) or

- demonstrations and symbolic representations (*17*).

Most researchers believe that frequent, consistent use of particulate animations is required for maximum benefits. For example, animations were used during each class for a two-week unit in the Williamson and Abraham (*16*) study. Although, Yezierski and Birk (*35*) used short duration animations in a single treatment, the authors recommend:

> "Particle-level animations should be used frequently in chemistry classrooms to help students visualize particle-level behavior." (pg 960)

Finally, students will quickly learn to disregard anything that is not valued by the instructor. Thus, if an instructor values the information in the animation, there will be assessment items or course points associated with the information. This can be achieved through participation points, points for the activity, homework, or examination questions. Sanger (*41*), who also agrees that particulate animations should be used consistently, proposes that molecular-level questions should be used in homework assignments and examinations. For those instructors not experienced with these conceptual questions about particles, such questions can be found in the homework problems and ancillary materials for most college-level textbooks. Additionally, conceptual questions asking students to predict particle behavior or give reasons for particle behavior can be found on-line from the *Journal of Chemical Education* and other sources. Some of these resources include the following:

- http://www.jce.divched.org/JCEDLib/QBank/collection/ConcepTests/
- http://people.brandeis.edu/~herzfeld/conceptests.html

In summary, it is a myth that instructors can simply show animated sequences intermittently throughout the course. Instead, particulate animations should be used consistently throughout the course. Additionally, instructors should consider using active learning strategies in conjunction with particulate animations. Finally, the material in animations should be assessed or count for some course points.

Myth #3: It doesn't matter when you use an animation in the lesson sequence.

Although frequent, consistent use of animations is recommended, many instructors have questions about when animations should be incorporated. Since it is suggested that animations be used in conjunction with other activities, should the particulate animation come first or last? Does it matter?

Velazquez-Marcano, et al. (*42*) investigated the benefits of using a demonstration and a particulate animation with second-semester, college chemistry students, including if there was a difference due to presentation order. As the investigators decided what variables to collect, they felt that certain characteristics of the students might play a role, so demographic data were collected. Students were asked about their gender, classification in college, grade point average, date of birth, and the number of chemistry and physics courses completed in both college and high school. Past physics courses were deemed to be important because the content in the study dealt with effusion/diffusion of fluids (gases and liquids). Since reasoning ability had been a factor in previous studies (e.g., (*16*)), investigators also asked students to take the Group Assessment of Logical Thinking (GALT). This test measures six types of logical thinking: conservation, control of variables, correlations, proportions, combinations, and probability (*43*, *44*). Finally, students were also asked "which teaching strategy would you say you learn best from?" Options were lecture only, laboratory, demonstration, and animations. All students were first shown two adjoining flasks and asked to predict what the system would look like when the valve between the two was opened. They were also asked to give the reason for their prediction. The prediction was chosen from seven multiple-choice options, while there were 10 multiple-choice options for the reason. All of the multiple-choice options were generated from pilot studies where students gave free responses. The choice options used in this study represented the most common answers given in the pilot studies.

The students were randomly assigned to one of two treatment groups by the computer. One group of 90 students first saw a video of a demonstration in which the valve was opened, followed by a particulate animation of the particles in the flasks (VA group). A second group of 81 students saw the particulate animation, followed by the demonstration video (AV group). After students gave a prediction and reason for the static picture of the flasks, they were next shown the first visualization (either the video or the animation, depending on the treatment group) and allowed to change their prediction and their reason. Finally, students were shown the second visualization (either the video or the animation, depending on the treatment group) and allowed to change their prediction and their reason. This sequence was repeated 3 times. Initially, students were given the scenario where NO2 gas was in the bottom flask and a vacuum was in the top flask (experiment 1). Next liquid water was in the bottom flask, with a vacuum in the top flask (experiment 2). Finally, there was NO_2 gas in the bottom flask and air in the top flask (experiment 3).

This paper did not analyze the responses from the ten options for the reason selected by students. Investigators discovered that analysis was almost impossible due to too many choice options and combinations. A later study was conducted to look at the reasons. After all the prediction data were analyzed, investigators found that there was no relationship between the correct predictions of what the flasks would look like when the valve was opened and any demographic factors (gender, reasoning ability, etc.). About 40% of the students, said they learned best from demonstration, about 33% said they learned best from lecture only, about 20% said they learned best with animations, and only about 3-5% preferred the laboratory. This data concerning teaching strategy preference is confounded by the uncertainty in how the students interpreted the response choices. Their learning preferences did not correlate to the correctness of predictions.

The most interesting result from the study was that considering all three experimental scenarios, the number of correct predictions significantly improved with the first visualization and significantly improved again with the second visualization. The order of the visualizations did not make a difference, but both were needed to gain the maximum benefit. This seems a bit funny, because when students in the VA group viewed the video demonstration of the brown NO_2 gas and the vacuum, they saw the final result where both flasks were equally colored. Why didn't the maximum number of students then change to the correct prediction? Significantly more students gave the correct prediction after the particulate animation. Investigators believe that even after seeing a video of the demonstration some students held tenaciously to their misconceptions when predicting the outcome and that for some students, the animation was required to provide an explanation of why the correct prediction should be accepted.

The animation alone was not enough to get the maximum benefit for students in the AV group. Investigators believe that the particulate animation alone was not enough to allow students to evoke macroscopic mental models to use in predicting the macroscopic outcome in the flasks. Again, significant improvement was found after the video demonstration was viewed. The study showed that both types of visualizations (a particulate animation and a macroscopic demonstration video) are needed, but the study did not show any advantage for order.

In a subsequent study by many of the same investigators (*45*) the choices for the predictions were limited to five (the top five chosen responses from the original seven) and the choices for the reason were limited to four (again the four responses most often given from the original ten). In addition students were allowed to type in a response if they did not like the four reasons provided. After students gave the initial prediction from the static drawing of the two flasks, they were shown the first visualization (either animation or video depending on the treatment) and afterwards were shown the correct prediction and asked to choose a reason (but not change their prediction as with the first study). Students were then shown the second visualization and asked to choose the reason for the action on the screen again. A different turn in this study was that students were next shown a third visualization, which was a side-by-side view of the first and second visualization, and were asked again to choose their reason. This sequence was repeated through the same three experimental scenarios as in the previous study (NO_2 and vacuum; liquid water and a vacuum; NO_2 and air).

The desired outcome was that students would choose particulate-level explanations of the phenomena. The VA group showed significant improvement when moving from the video to the animation, while the AV group showed a significant decrease in 2 of the 3 experiments as they moved from the animation to the video. This might suggest that students were only giving particulate answers after the animation. However, when the two groups were compared, the VA group (n=207) consistently gave significantly more particulate level responses that did the AV group (n= 249) whether you consider the groups after their first visualization, after both of their visualizations, or even after the animation, regardless of the difference in order. This study also showed that the combination view might have caused some confusion, as it produced a decline in particulate reasons chosen, perhaps allowing some students to revert back to their macroscopic explanations. It should be noted that the VA group, while it did decline with the combination view, was still significantly higher than the AV group The authors suggest further study into the combination views. This study showed that both demonstration videos and particulate animations are needed for students to choose the maximum particulate explanations of the phenomena, although putting them side-by-side did not lead to improvement in student responses. However, order does make a difference. The macroscopic view afforded by the demonstration videos followed by the particulate animations gave the best results.

Other researchers have called for macroscopic activities first followed by particulate ones, but few studies have been done in this area. For example: Russell, et al. (*17*), when discussing their computer program that contained macroscopic videos, particulate animations, a graphical display, and appropriate symbols, suggested that:

> "Some instructors, after showing first video only, will show video-animation, video-graph, and animation-graph to allow class discussion of the links between these representations and finally all views together to discuss the advantages and limitations of each representation." (p 332)

Tasker (*40*) describes his VisChem learning design as moving from macroscopic to particulate. He outlines the learning design as beginning with having the student observe a phenomenon, which is a macroscopic view. Next the student should be asked to describe and draw a molecular-level representation, discuss the drawings with peers, and view animations, all of which are particulate-level representations. He continues the design with asking the students to reflect on any differences with prior conceptions by comparing their drawings to the animation and to relate the two. Tasker finishes the learning design by assessing students' understanding of chemistry concepts to check for their adaptation of the views.

In summary of the literature presented behind this myth, it seems that there may be a preferred order of presentation. While more research needs to be done in the area, most agree that instructors should begin with the macroscopic view, then progress to the particulate and symbolic levels of representation. This order fits with Piagetian ideas that normal movement as we age is from concrete to more abstract thinking (*46*). The macroscopic activities like laboratories or demonstrations are more concrete in nature, while thinking about particle behavior is more abstract.

Myth #4: Computer-based visualizations are hard to find. (It takes a lot of time to find the animations that fit lectures.)

Some instructors believe that computer-based visualizations are difficult to find and that incorporating visualizations of any type into their teaching will require a lot of preparation time. These ideas are becoming more outdated as time passes. It is true that a certain amount a preparation time is required. After a 3-week summer workshop, Williamson, et al. (*47*) identified five factors when high school teachers tried to implement molecular visualizations in the classroom. These factors were both barriers when not present and facilitators when present.

These included Internet access, computer lab availability, adequate preparation time, technical support availability, and availability of administrative support.

In order to minimize preparation time, instructors must find the visualizations that match the learning goals they have for the class, and then instructors must try these visualizations in their classroom setting. It may be that instructors should incorporate visualizations into one unit of study at a time. The good news is that visualizations are now much easier to find through a number of sources, including some of the resources described below.

A number of publishers offer free ancillaries that include a CD or website collection of visualizations. Check with the representative of the company publishing your textbook. Often animations are available from the textbook CD or website as QuickTime or avi movies. On a number of CD's if you do not want to play the animations in the publishers' interface, the animations will be available in a file on the CD that is usually entitled "media" or something similar. Once you drag the desired animation to your desktop, you can rename it and insert it in a presentation or play it as a stand-alone movie.

The Internet is also a source of free animations. Simply using a search engine (Google, etc), an instructor can find animations for specific topics. Most animations can be downloaded onto a personal computer, others have to be played from the Internet, which will require Internet access in the classroom. Samples of the websites available are listed below. Collections of animations and other resources can be found at:

- http://www.merlot.org/merlot/index.htm. Multimedia Educational Resource for Learning and Online Teaching
- http://www.chemeddl.org/collections/index.html. Chemical Education Digital Library
- http://www.chem.iastate.edu/group/Greenbowe/sections/projectfolder/animationsindex.htm. Chemistry experiment simulations, tutorials and conceptual computer animations from Thomas Greenbowe's group at Iowa State
- http://genchem1.chem.okstate.edu/CCLI/Startup.html. Interactive animations to be used in inquiry activities from John Gelder and Michael Abraham
- http://www.learningdesigns.uow.edu.au/tools/info/T4/index.html. A collection by Roy Tasker's group

Sometimes an instructor cannot find a simple animation that he/she wants to show. In those cases, simple animations can be produced using the free ChemSense software, available at http://chemsense.org/. This program is easy to use and will export animations as QuickTime or XML movies. This option should be used after the other suggestions to find previously developed animations have not been fruitful. The animations produced with ChemSense are not 3-D or commercial quality, but they may still serve the instructor's purpose. Remember that ChemSense was developed to assess student understanding. There is commercial software available for developing animations, but these programs require a substantial investment of time to learn how to use them, however, your department may have a graphics staff person who can help.

There are so many visualizations available that often the problem is the time needed to find the ones that will work for your teaching goals. The best advice is to start with one unit of study, using the animations throughout the unit in conjunction with other activities. You can also consider working with a colleague to pool resources.

Myth #5: I can't use all types of visualizations, so I can't use them at all.

Instructors often find that some techniques just do not fit with their teaching style, so they discount all visualization techniques. Just because an instructor feels he/she cannot use role-playing, this does not preclude the use of other techniques. With the knowledge that it is best to use multiple techniques and that recurring use is needed for maximum benefits, instructors must choose the techniques that work best for them. It is important to match the concept you are trying to develop, your learning goals for the class, your teaching style, your class, and your facilities. For example, the lack of room for role-playing or lack of technology to project animations can limit the techniques that you can consider.

Myth #6: Visualization techniques don't cause misconceptions.

It is easy when endorsing something or someone to overlook flaws or problems. Some instructors enthusiastically use visualization techniques, but seem oblivious to their limitations or potential for harm. As chemists, we understand both the benefits and dangers involved in the use of acetone or benzene. As a result, chemists can avert problems involved with using these solvents. We must also be aware of the potential dangers with visualization techniques so we can avert those potential problems as well.

Animations have been found to be the source of misconceptions by a number of researchers (e.g., (*19*, *20*, *38*, *40*)). Concerns are that animations with simplified views may cause misconceptions, while those that are too elaborate may overwhelm students also causing misconceptions. This is seen if the instructor has not consistently used animations that are progressively complex. The use of gradually more complex animations allows students to build their visual literacy by gradually building mental models. By using some of the techniques that researchers have suggested in conjunction with animations, instructors can probe for student interpretation of animations in an effort to minimize misconceptions.

For example: Tasker (*40*) recommends using his VisChem learning design where the student moves from the macroscopic to describe and draw a molecular-level representation, discusses the drawings with peers, and then views animations. Further, he suggests using animations of the behavior of water, solid salt, and molten salt prior to using the animation for salt dissolving in water. This will build a student's visual literacy, whereas just showing the salt dissolving may be too complex for students to derive a sound meaning. For these animations, students would reflect on differences between their drawings and the animations, relating both of these to each other and to the macroscopic before they adapt their understanding.

Summary

Instructors can prudently use visualizations to help their students' understanding of particle behavior by helping the students form correct mental models of the phenomena of interest. The use of visualizations can take little class time, but should be used consistently and in conjunction with other activities, preferably active learning activities. The content from visualizations must be important enough to evaluate, and as a result, animations and the concepts they cover should be included in assessment activities. Animations are perhaps the most popular type of visualization technique used and can be readily found from publishers or on the web. It is best to begin a topic with activities that involve the macroscopic view of the chemical phenomena and then move to the particulate and symbolic. The instructor must choose the visualization techniques to incorporate in the classroom that best match his/her goals for the class, teaching style, facilities, students, and chemistry content. In choosing these techniques it is important to consider the limits of the visualizations and discuss these with the class. Hopefully, the information in this chapter has provided evidence to combat common misconceptions or myths concerning using visualizations in the classroom, so that instructors can knowledgeably integrate visualizations into their courses.

References

1. *Encarta World English Dictionary*, computer software within Microsoft Word 11.5.5; Microsoft Corporation, 1999.
2. Johnstone, A. J. *J. Chem. Educ.* **1993**, *70*, 701–705.
3. Abraham, M. R.; Williamson, V. M.; Westbrook, S. L. *J. Res. Sci. Teach.* **1994**, *31*, 147–165.
4. Haidar, A. H.; Abraham, M. R. *J. Res. Sci. Teach.* **1991**, *28*, 919–938.
5. Williamson, V. M.; Huffman, J.; Peck, M. L. *J. Sci. Educ.* **2004**, *81*, 891–896.
6. Williamson,V. M.; Jose, T. J. Using Visualization Techniques in Chemistry Teaching. In *Chemists' Guide to Effective Teaching: Volume II*; Pienta, N. J., Cooper, M. M., Greenbowe, T. J., Eds.; Prentice Hall: Upper Saddle River, NJ, 2009; pp 71−89.
7. Talley, L. H. *J. Res. Sci. Teach.* **1973**, *10*, 263–269.
8. Wu, H.; Shah, P. *Sci. Educ.* **2004**, *88* (3), 465–492.
9. Nicoll, G. *J. Chem. Educ.* **2003**, *80*, 205–213.
10. Friedel, A. W.; Gabel, D. L.; Samuel, J. *Sch. Sci. Math.* **1990**, *90*, 674–682.
11. Battino, R. *J. Chem. Educ.* **1979**, *56*, 39–41.
12. Lerman, Z. *J. Chem. Educ.* **1986**, *63*, 142–143.
13. Barnea, N.; Dori, Y. J. *J. Sci. Educ. Technol.* **1999**, *8*, 257–271.
14. Dori, Y. J.; Barak, M. *Educ. Technol. Soc.* **2001**, *4*, 61−74. http://www.ifets.info/journals/4_1/dori.html.

15. Bunce, D. M.; Gabel, D. *J. Res. Sci. Teach.* **2002**, *39*, 911–927.
16. Williamson, V. M.; Abraham, M. R. *J. Res. Sci. Teach.* **1995**, *32*, 521–534.
17. Russell, J.; Kozma, R.; Jones, T.; Wykoff, J.; Marx, N.; Davis, J. *J. Chem. Educ.* **1997**, *74*, 330–334.
18. Harrison, A. G.; Treagust, D. F. *Sch. Sci. Math.* **1998**, *98*, 420–429.
19. Kelly, R. M.; Jones, L. L. *J. Chem. Educ.* **2008**, *85*, 303–309.
20. Kelly, R. M.; Barrera, J. H.; Mohamed, S. C. *J. Chem. Educ.* **2010**, *87*, 113–118.
21. Milne, R. W. *J. Chem. Educ.* **1999**, *76*, 50–51.
22. Schank, P.; Kozma, R. *J. Comp. Math. Sci. Teach.* **2002**, *21*, 253–279.
23. Williamson, V. M.; Watson, J. Changes in students' spatial ability, equilibrium content knowledge, and attitudes through the use of storyboards and student-constructed computer animation in college general chemistry. Manuscript to be submitted for publication.
24. José, T. J.; Williamson, V. M. *J. Chem. Educ.* **2005**, *82*, 937–943.
25. Abraham, M. R.; Gelder, J. I.; Haines, K. *Chem. Educ.* **2001**, *6*, 307–308.
26. Hoffler, T. N.; Leutner, D. *Learn. Instr.* **2007**, *17*, 722–738.
27. *Visualization and Science Education*; Gilbert, J., Ed.; Springer: The Netherlands, 2005.
28. *Visualization: Theory and Practice in Science Education*; Gilbert, J., Reiner, M., Nakhleh, M., Eds.; Springer: The Netherlands, 2008.
29. *Multiple Representations in Chemical Education*; Gilbert, J., Treagust, D., Eds.; Springer: The Netherlands, 2009.
30. von Glasersfeld, E. *Radical Constructivism: A Way of Knowing and Learning*; Falmer: Washington, DC, 1995.
31. Johnson-Laird, P. N. Mental Models. In *Foundations of Cognitive Science*; Posner, M. I., Ed.; MIT Press: Cambridge, MA; 1989, pp 469−499.
32. Tobin, K.; Capie, W. *Educ. Psychol. Meas.* **1981**, *41*, 413–423.
33. Novick, S.; Nussbaum *J. Sci. Educ.* **1981**, *65*, 187–196.
34. Gabel, D. L.; Samuel, K. V.; Hunn, D. *J. Chem. Educ.* **1987**, *64*, 695–697.
35. Yezierski, E. J.; Birk, J. P. *J. Chem. Educ.* **2006**, *83*, 954–960.
36. Sanger, M. J.; Campbell, E; Felker, J; Spencer, C. *J. Chem. Educ.* **2007**, *84*, 875–879.
37. Burke, K. A.; Greenbowe, T. J.; Windschitl, M. A. *J. Chem. Educ.* **1998**, *75*, 1658–1661.
38. Kelly, R. M; Jones, L. L. *J. Sci. Educ. Technol.* **2007**, *16*, 413–429.
39. Abraham, M. R.; Gelder, J. I. Technology Based Inquiry Oriented Activities for Large Lecture Environments. In *Chemists' Guide to Effective Teaching: Volume II*; Pienta, N. J., Cooper, M. M., Greenbowe, T. J., Eds.; Prentice Hall: Upper Saddle River, NJ, 2009; pp 173−184.
40. Tasker, R. Using Multimedia to Visualize the Molecular Work: Educational Theory into Practice. In *Chemists' Guide to Effective Teaching: Volume II*; Pienta, N. J., Cooper, M. M., Greenbowe, T. J., Eds.; Prentice Hall: Upper Saddle River, NJ; 2005, pp 195−211.
41. Sanger, M. J. Computer Animations of Chemical Processes at the Molecular Level. In *Chemists' Guide to Effective Teaching: Volume II*; Pienta, N. J.,

Cooper, M. M., Greenbowe, T. J., Eds.; Prentice Hall: Upper Saddle River, N.J, 2009; pp 198–211.

42. Velazquez-Marcano, A; Williamson, V. M.; Ashkenazi, G.; Tasker, R.; Williamson, K. C. *J. Sci. Educ. Technol.* **2004**, *13*, 315–323.
43. Bunce, D. M.; Hutchinson, K. D. *J. Chem. Educ.* **1993**, *70*, 183–187.
44. Roadrangka, V.; Yeany, R. H.; Padilla, M. J. *J. Res. Sci. Teach.* **1985**, *22*, 743–759.
45. Williamson, V. M.; Lane, S.; Gilbreath, T.; Tasker, R.; Ashkenazi , G.; Williamson, K . C; MacFarlane, R. The effect of viewing order of macroscopic and particulate visualizations on students' particulate explanations. Manuscript in review.
46. Bodner, G. M. *J. Chem. Educ.* **1986**, *63*, 873–878.
47. Williamson, V. M.; Brown, L. M.; Peck, M. L.; Simpson, M. *TX Sci. Teach.* **2005**, *34*, 12–16.

Chapter 7

"These Kids Can't Do Inquiry," Another Urban Legend

Patrick L. Daubenmire,*,[1] Donald J. Wink,[2] and Adam Tarnoff[3]

[1]Department of Chemistry and Center for Science and Math Education, Loyola University Chicago, Chicago, Illinois 60660
[2]Departments of Chemistry and Learning Sciences, University of Illinois at Chicago (UIC), Chicago, Illinois 60607
[3]Department of Education, Adler Planetarium, Chicago, Illinois 60605
***E-mail: pdauben@luc.edu**

"These kids can't do inquiry." This claim is often made in reference to students who have a low level of academic engagement and/or who lack proficiency in one or more of the following basic skills: reading, math, graphing, logical problem solving, and collaborative teaming. To challenge this claim, we address a perception that "doing inquiry" means leveraging a student's *preexisting* inquiry skills in order to develop the students' content understanding. We claim, instead, that by utilizing a specifically designed and aligned system of supports, teachers can assist students in developing inquiry skills and content understanding *simultaneously*. In other words: "Students learn what teachers teach." To support this claim we share evidence from a design-based methodology within an urban reform initiative which supports the claim that teachers can and do shift toward using an inquiry-based instructional framework. In doing so, they discover that students, even those with low or delayed abilities, respond and demonstrate progress in skill acquisition and in learning science.

Context

The High School Transformation initiative was created in response to a growing dilemma within a large urban district, that less than 6% of its students were receiving a college degree (*1*). The District conducted a thorough analysis to determine reasons for this poor performance and found a several issues: Firstly, a wide variety of textbooks and/or curricula were used to teach core subject areas – English, math, and science. For example, among 47 high schools teaching biology, it was found that 30 different textbooks were being used (*2*). Sometimes different textbooks were being used for different sections of the same course within the same school. Secondly, many teachers were found to lack proficient content knowledge themselves and held low expectations for student performance. This led to instructional practices that rarely went beyond fact-based, procedural questions. Thirdly, District and school-level support for teachers was weak and did not target specific needs for student learning in this environment. Finally, school leadership was fragmented and infrequently focused on developing effective instructional practices.

In response to these findings and with the support of the Bill and Melinda Gates Foundation, the District developed the High School Transformation strategy. Six change levers were described in which the District needed to: (1) define excellence and raise expectations; (2) attract and develop great leaders; (3) foster a belief that the school is the unit of change; (4) attract and develop great teachers and provide them with holistic instructional supports; (5) broaden alterative pathways to graduation; and (6) ensure 9th grade preparedness. Within the "great teachers" lever (4), the District sought to enable highly effective and aligned instruction in troubled schools and to foster powerful teacher growth over time. To do this, the District recruited partner institutions to help. Each partner was asked to develop of a set of holistic, integrated curricula for grades 9-11, course tools, specific and aligned assessments, strategies for instructional coaching, and professional learning opportunities for teachers. The District named these systems Instructional Development Systems or IDS (Figure 1).

IDS partners supported grade 9-11 teachers in math, English, and science. Schools applied to receive this support and were able to choose a provider for each subject area. Implementing schools conducted a three-year phase-in, 9th grade support in their first year of implementation, 10th grade support in their second year of implementation, and 11th grade support in their third year of implementation. To date, three waves of high schools have participated. Across the three waves, 45 district high schools received support of an IDS for math, English, and science. The Loyola-UIC Science Inquiry to Build Content IDS was one science IDS option schools could select. This IDS provided support to 11 high schools in the District within the initiative.

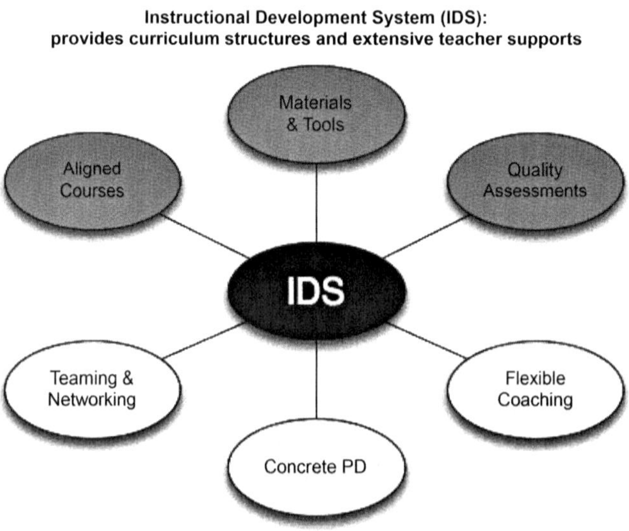

Figure 1. Diagram of IDS Support.

LUC-UIC IDS Logic Model

Our own subject of inquiry in program implementation was centered on changing and enhancing teachers' instructional practices using inquiry. We sought to build strong supports for teachers and infuse the program with opportunities for teachers who hold the myth that "these kids can't do inquiry," to revisit their reasons for this belief. To accomplish this, we developed a logic model (see Figure 2) for changing teacher practice.

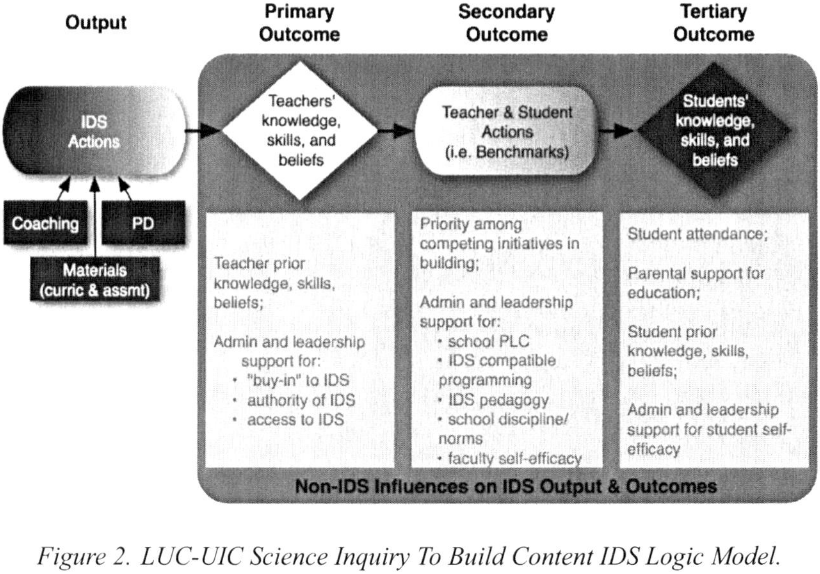

Figure 2. LUC-UIC Science Inquiry To Build Content IDS Logic Model.

If IDS materials, support, and actions successfully affected change in teachers' knowledge, skills, and beliefs for doing inquiry, then teachers' classroom actions would become more consistent with doing inquiry. The overall outcome was to affect change in students' knowledge, skills, and belief about doing inquiry.

This model is not without the influence of external, non-IDS factors. The school environment and culture, level and type of administrative influence and support, and students' level of participation (i.e. attendance) are among the factors that shape the outcomes.

The teacher and student outcomes which we determined would reflect a high implementation of the program are listed in our "Benchmarks of IDS Program" document (Appendix).

IDS Program Components

The curricular aspects of the program were built around inquiry-based, NSF-funded curricula: *Biology: A Human Approach* (*3*) in 9th grade, *Chemistry in the Community* (*4*) in 10 grade, and *Active Physics* (*5*) in 11th grade. The three courses were integrated through overarching themes within science (i.e. matter and energy, system and interactions, forces and motion, evolution, inquiry and technology, and nature of science) and scientific skills sets (i.e. asking testable questions, conducting scientific investigations, making claims based on evidence, interpreting and representing data, using mathematical reasoning, working together in groups, and communicating scientific information reliably and effectively.) Instructional cohesion across the three curricula was supported through a consistent framework of inquiry instruction based on the 5E approach (Table 1; (*6*)) as well as repeated use of the Science Writing Heuristic (Table 1; (*7–9*)) for classroom and laboratory activities.

The three-year sequence of science courses was aligned to local, state, and national standards. Funding also provided all the necessary equipment and consumable materials to conduct lessons and laboratory investigations. Each course was supported with 30 model lessons specifically designed for the IDS. These lessons articulated targeted learning objectives and alignment to standards, shared a detailed teaching plan and suggested pacing guide with recommended student activities and teacher actions. Content references and enhancement resources related to the science topic being studied and suggestions for providing differentiated instruction to accommodate students with special needs were also included. Resources for conducting embedded assessments plus sample rubrics were developed for each lesson. Formative assessments tied directly to each course were developed and administered throughout the school year. Finally, a comprehensive, content-specific summative exam was built for each area (biology, chemistry, and physics) across the IDS programs. Both the formative and summative instruments utilized multiple choice and free-response items which were aligned to state content and skill standards.

Table 1. The 5E Approach and Science Writing Heuristic in IDS

Beginning Ideas	Students **engage** by reading a short background passage or making an initial observation of a phenonemon and then formulating their own questions.
Tests	Students **explore** through specific tests that the instructor approves.
Observations	Students ask "what did you see? Are the results useful? If not, what can I do to change my procedure?"
Claims & Evidence	Students develop claims that **explain** their results based on specific evidence.
Reflection	Students **elaborate** by reflection on results, developing more detailed ideas in collaborations with other groups.
Conclusions & Meaning-making	Students **evaluate** what they have done in light of what is needed for their learning.

Program and curriculum-specific professional development was provided to teachers in a two-year sequence. A teacher new to using an IDS course received 60 hours of professional development. Thirty hours in the summer provided an orientation to program components, the teaching and learning philosophy and research-based practices as well as a thorough analysis of lessons and activities for the first unit of the year. The remaining 30 hours were divided across five, single-day sessions at strategic points in the academic calendar. These sessions targeted essential upcoming lessons and activities and shared various approaches to instruction and practices for analyzing student work. Teachers using an IDS course for a second, consecutive year received an additional 30 hours of professional development. These hours focused on developing a reflective practice and strategies for continuous improvement. These professional development hours included a user conference during which all teachers across IDS courses could present, share, and discuss best practices and needed goals for enriching students' learning within the program.

In cooperation with the curricular materials and professional development, teachers were provided with intensive instructional coaching. On a weekly basis, coaches provided specific instructional support to teachers. Several areas were targeted for coaching: sharing strategies for approaching content and lessons; co-teaching lessons, making observations and sharing feedback with teachers about instruction; collectively analyzing student work and planning instruction to address needs which surfaced; and assisting teachers in building professional learning communities (PLC) within their school teams (*10*).

Research Questions

To help us address the impact of program implementation, we developed two primary research questions:

1. By using an IDS-type intervention, do teachers change their instructional practices and strategies from using fact-based, procedural strategies (as found by the District's analysis) to using inquiry-based strategies?
2. If teachers do in fact shift to using more inquiry-based strategies, do they perceive that students show actions and responses consistent with inquiry skill and knowledge acquisition?

Methodology To Assess IDS Program Implementation

We chose to investigate these questions utilizing tools of design-based research. Design-based research, also known as design-based experiments, views curriculum design as proceeding through iterative cycles of design that are informed by the reflective analysis of systematically collected data on design implementation and outcomes. Curriculum designers analyze the collected data, reflecting upon the initial hypotheses of their design. If data indicate further refinement is necessary, designers then refine the design, implement the re-design, systematically collect data on the re-design, and reflect on the re-design (*11*). Since design efforts are implemented in real-world settings, design-based research endeavors recognize the importance of contextual factors, which results in research, design principals, and educational theories that are contextually sensitive (*12*).

One major goal of design-based research is theory development, with the iterative cycles of design providing researchers with the information and data to "gradually knit together a coherent theory that reflects their understanding of design." (*11*). With regard to theory generation, Edelson identified three types of theories that may emerge from design-based research: (1) domain theories, (2) design frameworks, and (3) design methodologies. Domain theories are descriptive theories that focus on generalizing some aspect of a design effort. These theories are focused on people and how they learn or on the learning environments and how they impact learning and teaching. The two major domain theories are context theories and outcome theories. Context theories attempt to generalize the contextual factors influencing the challenges and opportunities of a particular design effort. Outcome theories attempt to generalize the set of outcomes associated with a particular design intervention (*11*). Though both theories are necessarily interwoven, in addressing the myth, we focused on developing an outcome theory.

We contemplated the appropriate types of outcomes, or changes, that should be expected from a large-scale curriculum reform effort implemented in an urban school district. In other words, taking into account the environment in which the IDS program is implemented, what are reasonable indicators of change? Typically, outcomes are identified through problem analysis, or what is more commonly known as needs assessment which helps identify goals, challenges, constraints, and opportunities of a particular design context (*11*).

We engaged in problem analysis in several areas. Weekly, we analyzed coaches' field notes and used a framework to identify strengths, weaknesses, obstacles, and opportunities of program implementation (SWOO; (*13*)). Strengths helped identify what was working well and needed to continue to support the program. Weaknesses were identified as aspects that were not working well. Resources should be committed to resolve or fix these issues. Opportunities and obstacles were external (non-IDS) aspects of the program. Opportunities were viewed as ways to leverage other initiatives to support teachers' in doing inquiry. Examples of opportunities included school-wide literacy initiatives and district-wide initiatives to improve schools' behavior management and culture. Obstacles disrupted and hindered implementation. We focused efforts to isolate or minimize the impact of these influences. Examples included late delivery of needed textbooks and materials, school schedule and program changes continuing well into the academic year, and poor student attendance.

During professional development sessions we held focus groups and asked teachers to articulate their own version of a SWOO analysis. Annually we held an all-user focus group that also included the administration of two surveys: an online *Survey of Enacted Curriculum* (WCER) and teachers' reflections on their assessment of reaching programmatic benchmarks (*14*).

We planned this form of program analysis to assist in the development of an outcome theory for using an IDS type intervention and doing inquiry in schools with "these kids."

Who Are "These Kids?"

The claim, "these kids can't do inquiry," is often made in reference to students who have a low level of academic engagement and/or lack proficiency in one or more of the following basic skills: literacy, math, understanding and constructing graphs, logical problem solving, teaming and communicating effectively with peers (*15*, *16*). These kids are found in urban and rural areas, in low socio-economic status, below grade level in school, in special education and/or remedial courses, and within "failing" schools. As high as 40% of students at high schools using IDS programs were identified with some type of individual education plan (IEP). Essentially, these kids are found everywhere: in every school and in nearly every classroom.

What Does It Mean To "Do Inquiry?"

The Center for Science, Mathematics, and Engineering Education (CSMEE) describes fundamental abilities necessary to do scientific inquiry for grades 9-12 (*17*). High school students:

- Identify questions and concepts that guide scientific investigations.
- Design and conduct scientific investigations.
- Use technology and mathematics to improve investigations and communications.
- Formulate and revise scientific explanations and models using logic and evidence.
- Recognize and analyze alternative explanations and models.
- Communicate and defend a scientific argument.

Our program's specific benchmarks incorporate these fundamental abilities into the targets of implementation. When teachers examined these targeted abilities at the onset of program implementation, we encountered large support for the myth. The students that came to them as 9th graders had wide gaps in their set of skills and knowledge. They just simply could not do inquiry. Given the list of fundamental abilities, we could understand their claim. Even students who had good schooling prior to 9th grade might not be able demonstrate these fundamental abilities. However, due to the design of the IDS intervention and our hypothesis that "students learn what teachers teach," we wished to make the distinction between "can't" and "can't yet."

If students are not ready for grade level inquiry, then what are the implications for teaching these kids? Essentially, there are two categories of responses. The first follows the decision that *students are the limiting factors* to doing inquiry. Instructional planning in this category focuses on developing lectures and slide presentations to present concepts, which require students to memorize scientific terms, explanations, and models. These presentations can be accompanied with worksheets to drill students on term identification and/or to practice repeated calculations. Scientific investigations rely on verification of pre-determined results via "cookbook" labs which keep to one explanation or model to explain a scientific concept. Assigned readings summarize scientific arguments rather than having student construct their own ideas and arguments about scientific phenomena. This category clearly reflects the predominant direct, fact-based strategies used by teachers as identified by the District during its investigation.

In the other category of approaches, teachers may decide to adjust levels of inquiry to make inquiry-based instruction accessible to "these kids." Instruction can be designed to build content and skills in parallel by meeting students at their ability level and assisting them in reaching grade level abilities in inquiry. For example, if students are not ready to design and conduct their own scientific investigations, can they ask question about objects, organisms, and events in the environment; use appropriate tools and techniques to gather, analyze, and interpret data; and/or develop descriptions, explanations, predictions, and models using evidence (*17*)? This second category of approaches holds that *teachers, not students, are the limiting factor* (*15*, *16*, *18*). This premise informed our work. If teachers change what they do in the classroom, then students would change what they do.

Participants and Data Sources

Participation in the program involved 121 individual teachers in 11 different high schools who phased into implementation during a three-year period from academic year 2006-07 through 2008-09. At the completion of 2009, among teacher participants, there were 56 first-year users, 55 second-year users, and 10 third-year users. Nearly 6,500 students experienced lessons using this program across the three year period.

The data sources we selected to analyze implementation came from multiple sources. Program coaches provided a weekly coach's reflection following their work with teachers after an end-of-the-week implementation team meeting. Within these reflections, coaches were asked to describe what aspects of IDS lessons worked well; what was challenging about observed lessons; what evidence was present that teachers were working to attain program benchmarks in their classrooms; what were the common teacher problems and needs to be addressed; and how well were teacher teams working.

The pattern of school year professional development session allowed us to capture focus group feedback from teachers at different points during the year. First-year users participated in six sessions of professional development during the school year while second and third year users participated in three sessions. We used a framework of SWOO analysis to help with assessment of program implementation during these sessions.

Finally, we held an annual end-of-the-year reflection session with teachers. Among activities during these sessions, teachers provided feedback to two surveys. The first was a self-assessment of their progress in reaching programmatic benchmarks. For each of our benchmarks, teachers rated their own work, their students' work, and their professional learning community's work as "lacking," "emerging," or "practicing" for the particular benchmark. The second instrument was an online survey called the *Survey of Enacted Curriculum (SEC)*. The SEC data provides a set of measures for analyzing the relationship between the intended, enacted, and assessed curricula. Because teachers responded to this survey annually, we were able to collect repeated measures on some teachers who used the program in two or more years.

Implementation Analysis

Summary of Focus Group Feedback

Our analysis across feedback sessions revealed themes in the teachers' feedback. Three implementation team members assigned initial codes to these themes. Code descriptions were shared with other team members. Following discussion, we made any necessary coding adjustments and agreed to the following codes of SWOO feedback: overall classroom activities, organization of classroom activities, student literacy, resources used or provided by the program, other supplementary materials (provided or required), technology, timing, student engagement, student understanding, and student work. There was one additional code to which we assigned school/district-based administrative issues. Though this code had an overwhelming number of entries and clearly had an influence on the program, in this analysis we were primarily focused on responses directly related to teacher and student interactions with IDS program resources and materials, and did not include this codes in this analysis. Figure 3 shows the three codes with the highest frequencies from each section. An interesting note is that three categories appear multiple times. The multi-placement of these codes among the strengths, weaknesses, opportunities, and obstacles resulted from teachers assessing multiple parts of the materials. For example, the pacing guide which we provided to teachers on a biweekly basis was characterized as astrength based on its description of lessons and example lesson plans, but was also noted at times as a weakness for suggesting timing that was too fast for students.

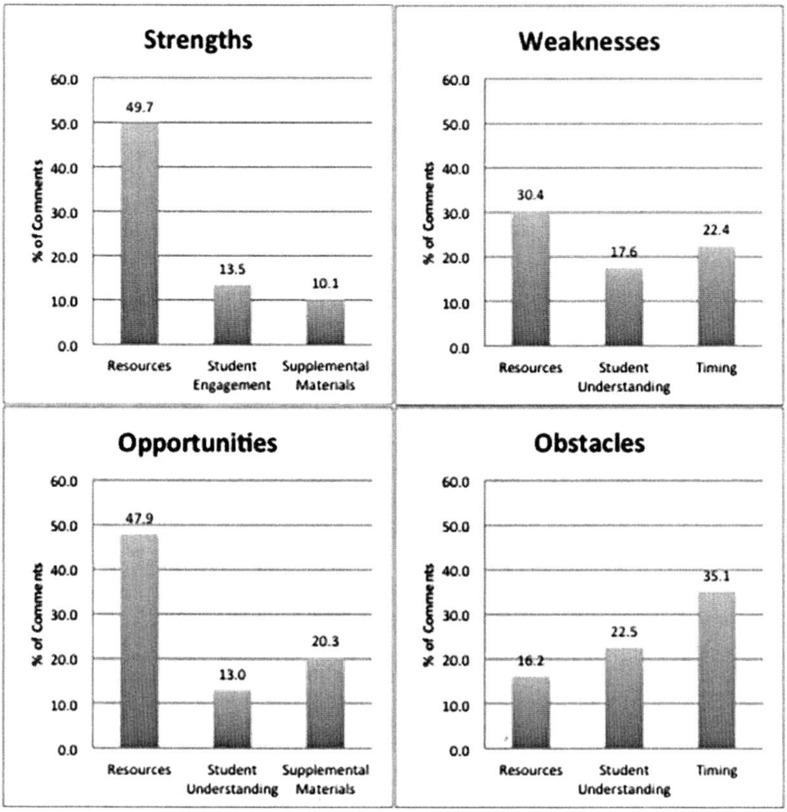

Figure 3. Code Summary of Comprehensive SWOO Analysis.

A look at teachers' specific comments from these SWOO analyses further reveals this multiple assignment of codes. Figure 4 shows a written response from a teacher focus group which places student discussion and responsible engagement as a strength, but with an arrow indicating that it was also a weakness. Teachers shared that focusing on students' discussion and engagement was something to be continued within the program. However, this group of teachers at the time of analysis (Fall 2009) thought that too much reliance on that strategy was a weakness.

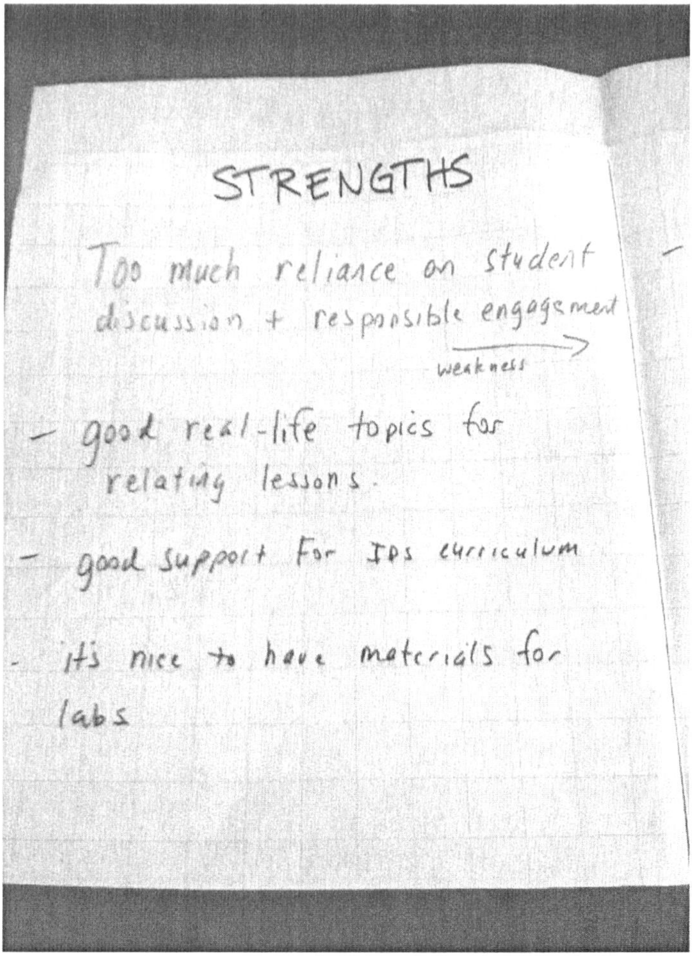

Figure 4. Sample Strengths Chart from Teachers Focus Groups.

Other sample comments provided on the next page exhibited this tension of emerging success of inquiry activities. In other words, teachers noted it was necessary to find the balance between the need for teacher direction and student readiness for autonomy in shaping their own ideas with regard to the content.

> "Students engage in activities but they need to improve in supporting their claims with evidence." (2nd-year user; Spring 2008)
> "Lab activities increase learning possibilities, but kids [sic] don't work in groups well." (2nd-year user, Fall 2009)
> "Instruction should foster more student reflection on what they have learned." (1st-year user; Fall 2009)
> "Helping students achieve at grade level remains a challenge." (1st-year user; Spring 2010)
> "With repeated entries into the inquiry cycle, students do eventually 'get it.' " (3rd-year user; Spring 2010)
> "Lab activities open ideas for my students, but I've got to get them to be patient." (2nd-year user; Fall 2009)

These feedback data were the first indicators that teachers using the IDS program might be utilizing more than a direct, procedural type of instruction and trying inquiry-based instruction. The tension about moving in the direction of inquiry approaches is revealed in the splitting of programmatic components in several SWOO categories. For example, activities that engage students in discussion about scientific ideas are described as strengths, but how much is given over to the student is an issue of student readiness. Though teachers describe students as still "can't yet," there are some degrees of inquiry that students can engage in according to teachers.

Teacher's Self Assessment of IDS Benchmarks

Along with focus group feedback, at the beginning and end of each year, teachers conducted a self-assessment of their progress toward reaching programmatic benchmarks (see Appendix), from lacking to emerging to practicing across academic year. Figure 5 shows a histogram for segments of growth reported by a percentage of teachers on two benchmarks from the beginning of the school year to the end of the school year. Additionally, these two examples show that a majority of teachers made progress toward a practicing level of these two benchmarks. It can also be observed that some teachers reported a decline in their progress toward program Benchmarks. In analyzing other data sources, the primary interferences were related to the non-IDS influences (described in our logical model) and not due to program components themselves.

From the beginning to the end of academic year 2008-09, teachers reported growth in 17 of the 20 Benchmarks. For student actions in their classrooms, the majority of teachers shared that students had improved levels from lacking to emerging for benchmarks 2, 3, 5, and 6. For 1, 4, and 8, teachers shared a level of improvement from emerging to practicing. Benchmark 7 did not show a noticeable progression or recession.

In the areas of "when I teach . . ." a majority of teachers shared that they moved from lacking to emerging in benchmarks 13 and 14 and shifted from emerging to practicing for benchmarks 9-12. Benchmark 16 showed the largest shift across the academic year, from lacking to practicing for the majority of teachers reporting. In these data, teachers have shared that their progression in reaching program benchmarks is accompanied by a progression, though not at the same rate, of student actions consistent with program benchmarks. These data indicate that both teachers and students can make classroom moves that are consistent with inquiry-based approaches.

Benchmark 9c:
When I teach I plan a strategy to identify and respond to students' initial understanding of the concepts.

% of Teacher in Each Category

Beginning of Year

	Lacking	Emerging	Practicing	
Chem	25.0	37.5	37.5	n=8
Bio	9.1	**68.2**	22.7	n=22
Total	13.3	**60.0**	26.7	n=30

End of Year

	Lacking	Emerging	Practicing	
Chem	0.0	88.9	11.1	n=9
Bio	0.0	48.0	**52.0**	n=25
Total	0.0	**58.8**	41.2	n=34

Histogram of Growth

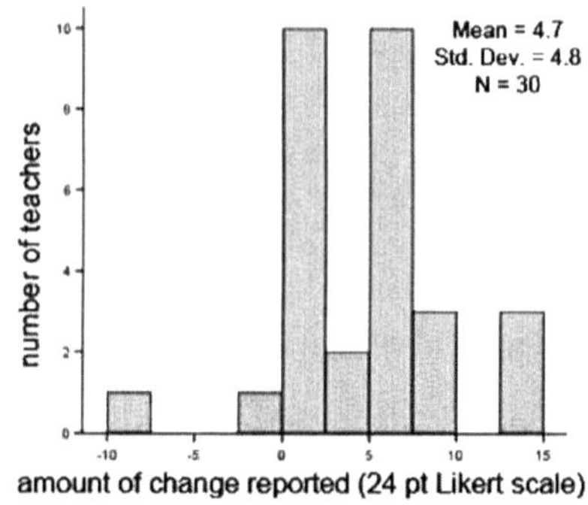

Benchmark 9d:
When I teach I plan a strategy to help students connect the lesson to the overall goals of my course and the "big ideas" of science.

% of Teacher in Each Category

Beginning of Year

	Lacking	Emerging	Practicing	
Chem	11.1	**77.8**	11.1	n=9
Bio	40.9	31.8	27.3	n=22
Total	32.3	45.2	22.6	n=31

End of Year

	Lacking	Emerging	Practicing	
Chem	0.0	30.0	**70.0**	n=10
Bio	0.0	24.0	**76.0**	n=25
Total	0.0	25.7	**74.3**	n=35

Histogram of Growth

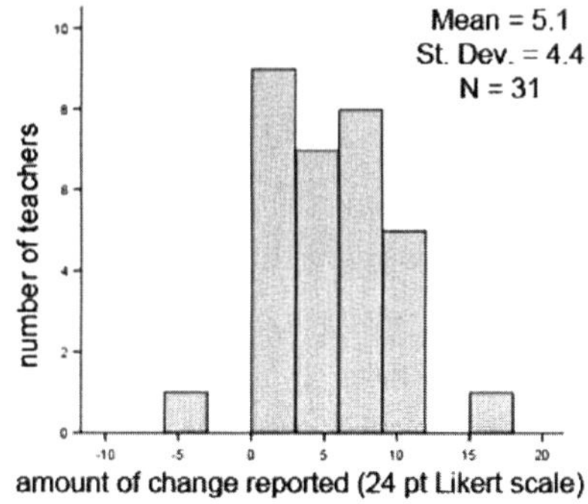

Figure 5. Sample set of Histograms of Growth on IDS Benchmarks.

Survey of Enacted Curriculum (SEC)

In addition to the self-assessment of IDS Benchmarks, annually we administered the Survey of Enacted Curriculum (*14*). This online survey asks teachers to report the percentage or proportion of time using particular instructional activities and how often they focus on certain concepts and skill development within a content area. In aggregated reports (individual teacher responses are disclosed only to the teacher), analyses can be made dealing with teachers' implementation of systematic, standard-based reform in science education (*19*).

When compared to all teachers within the state responding to this survey, teachers who had used the IDS for at least one year reported slightly more classroom time spent doing inquiry-based instructional activities. These activities included doing laboratory investigations, having students work in groups, asking students to maintain and reflect on their own science work. Additionally, IDS teachers with a minimum of one year experience using the IDS reported using whole-class lecture less than teachers throughout the state.

Extending our trend analysis, we looked at some aspects of teachers at various levels of experience within the program and within the state on inquiry-based laboratory activities (Figure 6). Examples of inquiry-based lab techniques included making predictions, designing experiments to solve a scientific question, making observations, and analyzing and interpreting scientific data. Greater percentages of time using inquiry-based laboratory techniques when compared to teachers in the state were not recognized until IDS teachers had spent at least two using the program.

In areas of inquiry-related instructional practices, we were able to analyze trends between teachers implementing the IDS program for only one year and teachers who implemented the IDS program for two or more consecutive years (multi-year users). Multi-year users reported that they were more prepared to manage a class of students using hands-on laboratory activities and to take into account students' prior conceptions when planning lessons than first-year users (~40% compared to ~20%). Furthermore, multi-year users employed active learning, performing laboratory procedures, and making connections among concepts and content some of the time (~30%). First-years users reported using the same practices a little of the time (~15%).

Given the responses to the SEC, as a second self-report indicator, teachers, upon repeated use of this IDS, report using techniques that move away from direct, fact-based instruction and toward using inquiry-based practices in the classroom.

The Adaptive Practice of IDS Coaching

Our coaching approach utilized a blended content-focused and strengths-based model (*20*, *21*). In order to select effective coaching strategies for optimizing the use of inquiry-based approaches, it is essential to understand a teacher's beliefs and attitudes, their knowledge of the content, their natural preferences for processing information and making decisions, and the influence of the school environment and culture. To apply this model, coaches worked regularly with teachers in their own school environments by:

- discussing and investigating approaches for assisting both new and experienced teachers with instruction and the curriculum materials,
- planning and observing instruction as well as co-teaching when asked or needed,
- discussing students' learning needs in particular areas and identifying or developing well-supported solutions to address those needs,
- studying and evaluating tools of assessment, and
- examining and utilizing effective skills and tools for mentoring and guiding teachers in forming professional learning communities.

During implementation, coaches collected field notes during each school visit. Weekly, the support team of the IDS, including coaches, met to assess progress of implementation. Field notes captured coaches' work as they co-planned with teachers, observed classroom implementation, analyzed student work and assessments, and engaged and guided school-based professional learning communities around using the IDS. Besides applying our own SWOO analysis each week, coaches provided a weekly reflection for teachers focused on aspects of fidelity of implementation. In each reflection, coaches shared their observations and insights about teachers' implementation as related to program benchmarks. At the conclusion of each reflection, coaches shared and collaboratively built strategies for addressing needs in upcoming weeks.

Statements from coach reflections provided the third set of indicators for teacher progress in using more inquiry-based approaches as well as observations of student behaviors in response to any changing teacher practices. (Teacher pseudonyms are used below in order to protect teacher participants' privacy.)

In working with a second-year user, a coach reflected, "Pat is quite proficient in conducting lessons according to the inquiry model; a noticeable improvement has been observed in expecting students to follow established norms." (Week 3, Fall 2008)

Reporting on a laboratory investigation, a coach shared, "the lab [this week] was enacted in a true inquiry manner. In fact students generated ideas for future exploration and possible science fair projects." (Week 2, Fall 2008)

Coaches also observed the tension, as described earlier, over doing inquiry in classrooms, "Students were engaged in lessons . . . However, Reese's modification of the lesson took away the exploratory feature of the initial investigation and students did not work cooperatively on any aspect of the lesson." (Week 15, 2008)

"Teachers did agree that their instruction was too much teacher-centered. I suggested that they look for ways to make it more student-centered and shared some approaches for doing so." (Week 12, Fall 2009)

Classroom observations also revealed student behaviors, "Students worked in teams to construct models of DNA. They enjoyed the DNA bead models" (Week 30, Spring 2010)

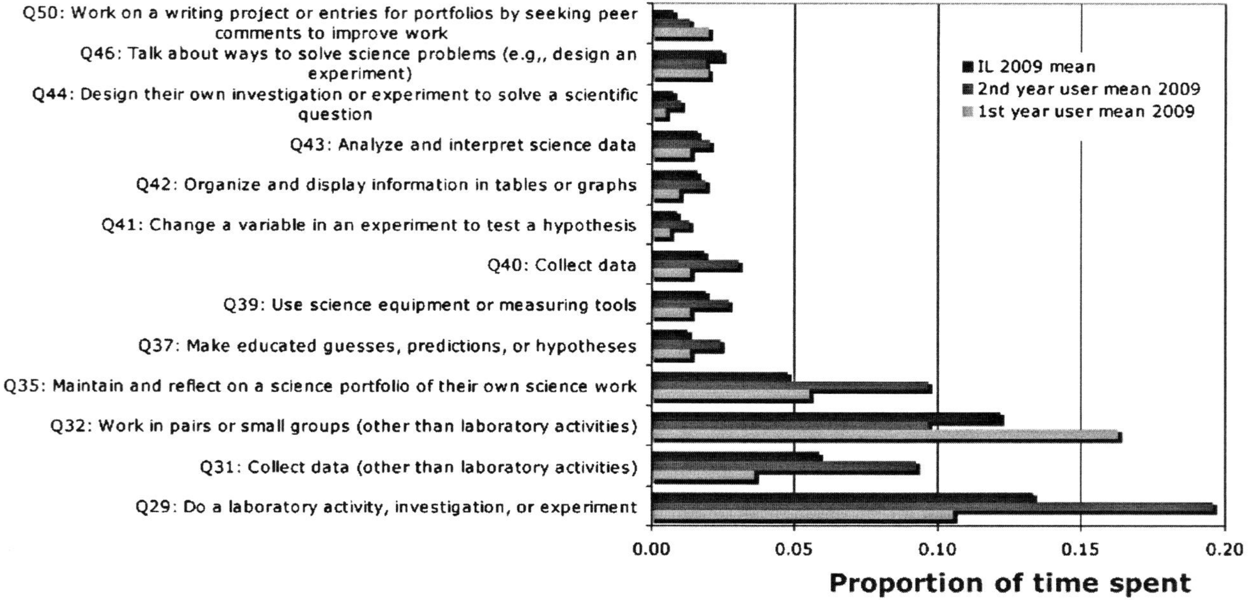

Figure 6. Sample School Teams Comparison of Proportion of Time Spent on Inquiry-Based Activities.

"I observed more students participating in discussions in class and taking more ownership of their own learning." (Week 8, Fall 2009)

Teacher action was also tied to student behaviors, "If Terry is consistent with classroom norms, students should start to respond positively to her soon." (Week 05, Fall 2009)

"Teachers are trying to devise strategies to differentiate instruction" (Week 15, Fall 2009)

These examples stem from a comprehensive set of reflections in which coaches describe patterns of development in teachers' use of inquiry practices and subsequent student actions when teachers used inquiry approaches. Table 2 organizes a set of exemplary cases from coach reflections in a contrast table in order to understand level of teacher practice (*22*).

Table 2. Contrast Table: Exemplary Cases Showing Levels of Teachers Inquiry-Based Approaches When Using an IDS

Aspect of Inquiry	*Ann (3rd-yr user)*	*Pam (2nd-yr user)*	*Timothy (2nd-yr user)*	*Laura (1st-yr user)*	*Greg (1st-yr user)*
Classroom Management	skilled	Skilled	emerging	practicing	lacking
Use of open-ended questioning	skilled	Emerging	lacking	emerging	lacking
Consistent structure of classroom inquiry (i.e. use of 5E model)	skilled	Practic-ing	emerging	emerging	lacking
Use of assessments to inform instruction	practicing	Practic-ing	emerging	lacking	lacking
Observed level of student behaviors reflecting use of inquiry	practicing (mostly at grade level inquiry)	practic-ing (pro-gressing toward grade level)	emerging	emerging (but tentative)	lacking

Coach reflections acknowledge that changing practice toward using inquiry is a process thatneeds time. "I didn't really like this program at first. I didn't see the connections. Now I see how helpful it is to students' learning. It is just hard work." (Teacher Response shared in Coach Reflection, March 2010). Even in the early use of inquiry instruction, teachers themselves do recognize changes to their practice as well as changes in the responses of their students when using inquiry.

Findings: Students Can Do Inquiry, If Teachers Do Inquiry

The set of data sources provide multiple perspectives for answering whether or not teachers can change practices from fact-based, direct instruction to inquiry instruction when supported by the IDS. The compiled SWOO analysis revealed the tension many participating teachers experience when using inquiry. The duplicate entries of codes in various SWOO categories showed that teachers find some features of inquiry, e.g. student engagement, to be strengths and opportunities, but too much engagement can be a weakness or obstacle, if students are poorly disciplined.

Teachers self-reports on two surveys demonstrated a perception by a majority of teachers that they did progress along the lacking to emerging to practicing continuum of inquiry instruction (Self-assessment survey data). They also articulated more frequent use of inquiry-based activities during multi-year use of the IDS (SEC data). They concurrently perceived growth, though delayed compared to their own, in student behaviors that reflected inquiry (Self-assessment survey data).

Finally, coaches' reflections concurred that teacher moves in the classroom affected how students responded. These teacher moves could be characterized within the inquiry instructional framework and at levels of implementation as well, from lacking to skilled. The progression to higher levels of implementation was accompanied by student actions that demonstrated parallel growth in using inquiry methods. The convergent results of these data sources led us to identify stages of use of inquiry teaching when supported by the Science Inquiry to Build Content IDS (Table 3).

Just as there are student levels for doing inquiry, teachers' adoption of inquiry-based instructional practices has stages of implementation. At each phase, students' opportunities to learn science content is shaped in a particular way toward higher levels of "doing inquiry." In other words, these kids *can* do inquiry. When instruction is limited to basic skills, students are limited to basic competency. When given the opportunity to learn complex inquiry skills, students do respond. With support, teacher practice changes gradually as teachers develop new knowledge, skills, and beliefs about inquiry-based teaching and learning.

Table 3. Stages of Implementing Inquiry Instruction with IDS

When teachers learn to . . .,	*students . . .*
Stage I – Emerging	
▪ manage class time and student behavior during lab experiments and teacher-directed group work; ▪ ask open-ended questions about observed patterns (e.g. how are these variables related?); ▪ develop classroom norms and routines to deliver consistent, uniform instruction from one day to the next, and; ▪ reflect on student learning outcomes demonstrated by immediate assessments (exit tickets, oral participation, etc.).	▪ work in cooperative groups to follow explicit directions and procedures+; ▪ use lab equipment to conduct experiments+; ▪ create data tables and written records of procedures+; ▪ recognize patterns and trends in data+, and; ▪ respond verbally to open ended questions+.
Stage II – Practicing	
▪ manage class time and student behavior during class discussions; ▪ ask open-ended questions about scientific concepts (e.g. how can we explain this relationship?); ▪ modify and adapt instruction based on student ability and prior knowledge, and; ▪ reflect on student learning outcomes demonstrated by close assessments (i.e. lab reports, analysis questions, final projects).	▪ work in cooperative groups to complete projects and produce presentations +; ▪ ask testable questions and design simple investigations*; ▪ create multiple representations of their own data (table, graph, equation, text) +; ▪ explain trends and patterns using scientific principles*, and; ▪ make written claims supported with evidence +.
Stage III – Skilled	
▪ manage class time and student behavior during student-directed group work; ▪ ask open-ended questions about reflective and metacognitive processes (e.g. how did you reach that conclusion?); ▪ help students to ask and answer their own open ended questions, and; ▪ reflect on student learning outcomes demonstrated by proximal assessments (e.g. tests aligned to state standards).	▪ work in cooperative groups to provide each other with peer-to-peer assessment and instruction*; ▪ design experiments with multiple variables and account for experimental errors in data*; ▪ relate data from more than one dataset*; ▪ evaluate the merits of competing claims*, and; ▪ reflect in writing about what they have learned and how their knowledge has changed*.
+ - approaching grade level inquiry; * - grade level inquiry	

Conclusion: A Design-Based Outcome Theory

In answering our research questions, the multiple perspectives of data sources indicate that teachers, when using the IDS, move into stages of inquiry implementation and away from direct, fact-based instruction. This is accompanied by an observable change in teachers' perceptions of students' ability to use inquiry practices. The culmination of this analysis led us to articulate a design-based outcome theory, which was the target for this particular experimental design. Figure 7 articulates this theoretical model.

The transition from "kids can't" to "kids can" is related to teacher willingness to adopt new strategies, such as inquiry-based instruction within IDS. The subtext of "these kids can't do inquiry" may actually be, "I am not sure I am yet ready to teach with inquiry." For the teachers who take the risk and begin to use inquiry, they find a realm that differs from one in which students can't because students are limiting, and find one in which students can because teachers are the limiting factor, not the students. In this new realm, ultimately, inquiry is possible and students can learn what teachers teach.

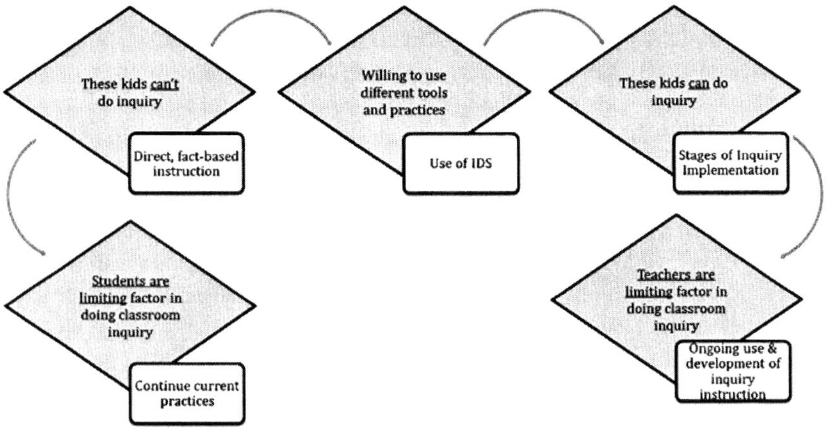

Figure 7. A Design-Based Outcome Theory Model for Using IDS to Implement Inquiry Instruction.

Appendix 1
Self-Assessment of Benchmarks of the IDS Program

Directions: Check one box per row. Please respond for only one subject (bio, chem, or phys).

Name: ______________ **School:** ______________

	In my classroom…	Lacking	Emerging	Practicing
	My students have input and adhere to a written set of classroom norms (group norms) for academic performance and classroom behaviors including clear references to:			
1a	○ active engagement in classroom activities.	□	□	□
1b	○ respectful collaboration with others.	□	□	□
1c	○ a responsibility for individual learning and the learning of others.	□	□	□
1d	○ a safe, secure environment for expressing ideas.	□	□	□

	In my classroom…	Lacking	Emerging	Practicing
2	**My students** can articulate learning goals and performance expectations.	□□□□□□□□	□□□□□□□□	□□□□□□□□
3	**My students** are explicitly aware of the scientific skills available and under development in their work, which include: interacting in and with groups of peers, communicating scientifically, designing and conducting scientific investigations, interpreting and representing data, and evaluating models, inferences, and experimental results.	□□□□□□□□	□□□□□□□□	□□□□□□□□
	My students participate in cycles of inquiry when they:			
4a	○ engage by generating questions and ideas and identifying personal theories about natural phenomena.	□□□□□□□□	□□□□□□□□	□□□□□□□□
4b	○ explore ideas through the formulation of testable questions and data-based investigations.	□□□□□□□□	□□□□□□□□	□□□□□□□□

	In my classroom…	Lacking	Emerging	Practicing
4c	○ explain by making claims based on patterns in the data and/or the support of relevant, accepted concepts.	□□□□□□□□	□□□□□□□□	□□□□□□□□
4d	○ elaborate through reflection on the meaning of the stated claims to their initial, personal theories.	□□□□□□□□	□□□□□□□□	□□□□□□□□
4e	○ evaluate their understanding by comparing it to the understanding of others and the accepted theories of experts.	□□□□□□□□	□□□□□□□□	□□□□□□□□
5	**My students** utilize literacy strategies when encountering and interacting with readings.	□□□□□□□□	□□□□□□□□	□□□□□□□□
	When I teach…	**Lacking**	**Emerging**	**Practicing**
6	**I** set high, consistent, and achievable expectations for academic, individual, and social actions.	□□□□□□□□	□□□□□□□□	□□□□□□□□
7	**I** guide development of a written set of classroom norms (group norms) for academic performance and student and teacher behaviors.	□□□□□□□□	□□□□□□□□	□□□□□□□□

	When I teach...	Lacking	Emerging	Practicing
8	I continuously monitor and provide reminders to students about classroom norms.	□□□□□□□□	□□□□□□□□	□□□□□□□□
	I plan a strategy to:			
9a	○ identify the most important concept or skill students should learn.	□□□□□□□□	□□□□□□□□	□□□□□□□□
9b	○ review the fundamental science content I need to understand before teaching the lesson.	□□□□□□□□	□□□□□□□□	□□□□□□□□
9c	○ identify and respond to students' initial understanding of the concepts.	□□□□□□□□	□□□□□□□□	□□□□□□□□
9d	○ help students connect the lesson to the overall goals of my course and the "big ideas" of science.	□□□□□□□□	□□□□□□□□	□□□□□□□□
9e	○ differentiated instruction to meet the needs of all students in my classroom.	□□□□□□□□	□□□□□□□□	□□□□□□□□

	When I teach...	Lacking	Emerging	Practicing
9f	○ assess each student's level of mastery of the primary learning objective.	☐	☐	☐
10	I articulate to students daily learning goals and performance expectations and engage with students in addressing the "so what" point of lessons and topics.	☐	☐	☐
11	I use technology (e.g. PASCO GLXs and sensors) where appropriate, as part of students' data-based investigations.	☐	☐	☐
12	I assist students in summarizing the salient points of each lesson.	☐	☐	☐
13	I individually reflect on how to continuously improve my instruction.	☐	☐	☐
14	I reflect with peers on student data and jointly develop improved ways to address pervasive student partial knowledge or misconceptions.	☐	☐	☐

	During course planning time with my colleagues...	Lacking	Emerging	Practicing
15	**We** establish and continuously monitor team norms.	□□□□□□□□	□□□□□□□□	□□□□□□□□
16	**We** meet 2-3 times weekly	□□□□□□□□	□□□□□□□□	□□□□□□□□
17	**We** jointly develop meeting agendas and the work of the planning time.	□□□□□□□□	□□□□□□□□	□□□□□□□□
18	**We** delegate tasks to individuals with deadlines.	□□□□□□□□	□□□□□□□□	□□□□□□□□
19	**We** consistently examine student work.	□□□□□□□□	□□□□□□□□	□□□□□□□□
20	**We** focus on classroom instruction.	□□□□□□□□	□□□□□□□□	□□□□□□□□
21	**We** support an environment conducive to risk-free discussion and healthy collegial debate.	□□□□□□□□	□□□□□□□□	□□□□□□□□
22	**We** analyze and evaluate student work collectively using the Tuning Protocol.	□□□□□□□□	□□□□□□□□	□□□□□□□□
23	**We** collaborate and consider alternatives when solving problems.	□□□□□□□□	□□□□□□□□	□□□□□□□□
24	**We** share particular teaching strategies and developed instructional tools.	□□□□□□□□	□□□□□□□□	□□□□□□□□

	During course planning time with my colleagues...	Lacking	Emerging	Practicing
25	**We** assume shared responsibility for materials	□□□□□□□□	□□□□□□□□	□□□□□□□□
26	**We** celebrate achievement and keep working for improvement.	□□□□□□□□	□□□□□□□□	□□□□□□□□

References

1. Cohen, J. S.; Little, D. Out of 100 Chicago Public School Freshman, Six Will Get a College Degree. *Chicago Tribune*, April 21, 2006, p 1.
2. *Annual Evaluation Report*; Chicago Math and Science Initiative (CSMI), 2003.
3. *BSCS Biology: A Human Approach*, 3rd ed.; Kendall Hunt Publishing: Dubuque, IA, 2006.
4. *Chemistry in the Community*, 5th ed.; American Chemical Society, W. H. Freeman: New York, 2006.
5. It's about Time. *Active Physics*, 3rd ed.; Herff Jones: Westerville, OH, 2009.
6. Bybee, R. *Achieving Scientific Literacy: From Purposes to Practices*; Heinemann Educational Books: Portsmouth, NH, 1997.
7. Keys, C. W.; Hand, B.; Prain, V; Collins, S. Using the science writing heuristic as a tool for learning from laboratory investigations in secondary science. *J. Res. Sci. Teach.* **1999**, *36* (10), 1065–1084.
8. Burke, K. A.; Greenbowe, T.; Hand, B. Implementing the science writing heuristic in the chemistry laboratory. *J. Chem. Ed.* **2006**, *83* (7), 1032.
9. Akkus, R.; Gunel, M.; Hand, B. Comparing an inquiry based approach known as the science writing heuristic to traditional science teaching practices: Are there differences? *Int. J. Sci. Ed.* **2007**, *29* (14), 1745–1765.
10. DuFour, R.; Eaker, R.; DuFour, R. *Revisiting Professional Learning Communities at Work: New Insight for Improving Schools*; Solutions Tree Press: Bloomington, IN, 2008.
11. Edelson, D. C. Design research: What we learn when we engage in design. *J. Learn. Sci.* **2002**, *11* (1), 105–121.
12. MacDonald, R. J. Professional development for information communication technology integration: Identifying and supporting a community of practice through design-based research. *J. Res. Tech. Ed.* **2008**, *40* (4), 429–445.
13. Ansoff, H. I. *Corporate Strategy*, revised ed.; Penguin Books: New York, 1987.
14. Wisconsin Center for Educational Research (WCER). Survey of Enacted Curriculum. www.seconline.org (accessed July 7, 2011).
15. Houltz, B. *Teaching Science Today*; Shell Education: Huntington Beach, CA, 2008.

16. Llewellyn, D. *Inquiry Within, Implementing Inquiry Based Science Standards*; Corwin Press: Thousand Oaks, CA, 2002.
17. Center for Science, Math, and Engineering Education. *Inquiry and the National Science Education Standards: A Guide for Teaching and Learning*. National Academy Press: Washington, DC, 2000.
18. Wendell, R. Dispelling the myths of inquiry teaching. *Clearing House* **1973**, *48* (1), 24–28.
19. Blank, R. K.; Porter, A. C.; Smithson, J. New Tool for Analyzing Curriculum, Teaching, and Standards in Mathematics and Science: Results from the Survey of Enacted Curriculum, Final Report; Council of Chief State School Officers, Washington, DC, 2001.
20. Staub, F. C.; West, L.; Bickel, D. What is Content-Focused Coaching? In *Content-Focused Coaching: Transforming Mathematics Lessons*; West, L., Staub, F. C. , Eds.; Heinemann: Portsmouth, NH, 2003; pp 1−17.
21. Kise, J. A. G. *Differentiated Coaching: A Framework for Helping Teachers Change*; Corwin Press: Thousand Oaks, CA , 2006.
22. Miles, M. B.; Huberman, A. M. *Qualitative Data Analysis: An Expanded Sourcebook*, 2nd ed.; Sage Publications: Thousand Oaks, CA, 1994.

Chapter 8

Do Students Learn Chemistry Content in Accelerated Summer Courses?

Mildred V. Hall,[1] Linda A. Wilson,[2] and Michael J. Sanger[*,2]

[1]Clark State Community College, Springfield, OH 45505
[2]Department of Chemistry, Middle Tennessee State University, Murfreesboro, TN 37132
[*]E-mail: mjsanger@mtsu.edu

There is a common perception among instructors and academic advisors that students are not as successful at learning chemistry content when enrolled in accelerated summer courses compared to the traditional 15-week fall or spring courses. This study, involving a secondary analysis of an existing data set, determined whether course format (three-week accelerated vs. 15-week traditional), students' academic experience (years of college credit taken), life experience (years since graduating from high school), and gender had an effect on students' chemistry content knowledge. This study showed that students enrolled in the accelerated course learned more chemistry content than students enrolled in the traditional course, and that students who had graduated from high school more than five years before taking the course learned more chemistry content than students who had graduated from high school less than five years before taking the course. Implications of this study and the limitations based on the use of existing data are also discussed.

The Myth—Student Learning in Accelerated Courses

Most readers will have heard this myth; some may even have said it themselves. There is an assumption that students are not as successful in learning chemistry content in time-compressed courses typically taught in the summer than they would be if they had taken the longer, traditional courses taught in the fall or spring. In informal discussions with faculty and staff members who advise students at one of our universities, the corresponding author asked them how they advised students with respect to traditional-length versus time-compressed (accelerated) chemistry courses. Several of them said they frequently advise students that they would be more successful in learning chemistry content in the traditional-length courses instead of the accelerated courses. But is this really the case?

Existing Literature on Accelerated Courses

Accelerated courses, also called *intensive courses*, have been extensively studied and several studies have compared student success in accelerated courses compared to traditional-length courses (*1–17*), including metaanalyses of multiple experiments (*7–9*, *15*). Although these results are mixed, most of these studies either showed that students' performance in accelerated courses was superior to those of students enrolled in traditional courses (*1–3*, *5–7*, *10*, *11*, *13*, *15*, *16*) or that student performance in the two course formats were equivalent (*4*, *7*, *14*, *15*, *17*). Several of these studies also included other variables that may affect student learning. These variables included the students' year in college/academic experience, age/life experience, and gender. Two studies showed that prior academic experience did not affect student learning (*8*, *13*) but another study showed that prior academic experience did lead to improved learning (*5*). Geltner and Logan (*7*) reported that students 17 and under performed better than any other age groups, but for students from age 18 and above student learning increased steadily as the students' age increased. The results for gender are mixed; one study reported no effect based on gender (*17*), one study found that male students outperformed female students (*11*), and another study found that female students outperformed male students (*13*).

Research Hypotheses

As part of her teaching assignments, the second author routinely taught a traditional-length (15-week) introductory chemistry course in the fall and an accelerated (3-week) version of the same course in the summer. Based on her teaching experiences, she felt that the students in the accelerated course learned more chemistry content than the students in the traditional course. This author collected data for students enrolled in these courses for four years, including course-related scores (test scores, final examination scores, course grades, etc.) and student records information (ACT scores, GPA, total college credit hours, years since high school graduation, etc.).

The research hypotheses for this study, developed by the first author as part of her doctoral dissertation (*18*), are:

1. Students in accelerated courses will develop better chemistry content knowledge than students in the traditional-length versions of the same course.
2. Students with greater academic experience will develop better chemistry content knowledge than students with less academic experience.
3. Students with greater life experience will develop better chemistry content knowledge than students with less life experience.
4. Males and females will develop different chemistry content knowledge.

In this study, students' chemistry content knowledge was measured using a 24-question multiple-choice posttest and the same 100-question multiple-choice final examination (containing the 24 posttest questions and 76 others) written by the instructor. Another paper comparing student success in these classes, as measured by students' total points and final grades earned, has been submitted to the *Journal of Chemical Education* (*19*). While the study described in this paper focuses solely on students' chemistry content knowledge, the definition of student success used in the other paper includes scores that also measure students' commitment (motivation) to attend and participate in the classroom and laboratory activities.

Methods

Subjects

The sample consisted of 130 students who were enrolled in 15-week sections of an introductory general chemistry course taught by the second author from Fall 2000 to Fall 2003 and 88 students who were enrolled in 3-week sections of the same course taught by the same instructor from Summer 2001 to Summer 2004. In addition to students' scores on the pretest, posttest, and final examination, we also had access to information from the students' academic record including ACT scores (for about 75% of the students), GPA and the number of college credit hours taken at the time of the course, the year that these students graduated from high school, and their gender. These data were analyzed using an analysis of covariance (ANCOVA).

Dependent Variables

Originally, the posttest score was used as the dependent variable in the ANCOVA analysis because it represents students' content knowledge of the 24 chemistry questions written by the instructor. However, we decided to perform another ANCOVA analysis using the 100-question final examination score with the 24 pretest questions removed (scored out of 76 points) as the dependent variable to measure students' chemistry content knowledge.

Covariates

The pretest, which was administered to the students on the first day of classes, and the student's GPA at the time of the course were used as covariates to correct for individual student differences. Since we were missing ACT scores for about one-fourth of the sample population, we decided not to use ACT scores as a covariate since it would reduce the sample size. A previous analysis of student succes (based on total points and students' grades) from this sample showed the same statistical results whether ACT scores were used as a covariate (and one-fourth of the sample was lost) or were excluded as a covariate (*19*).

Independent Variables

Based on the existing literature studies, we chose to use four measures as independent variables. These measures included course format (traditional or accelerated), academic experience, life experience, and gender (male or female). Academic experience was defined as the number of years of college credit taken by the subjects before enrolling in this course. This variable was divided into five categories: Less than one year (0-30 credit hours), one to two years (31-60 credit hours), two to three years (61-90 credit hours), three to four years (91-120 credit hours), and more than four years (120+ credit hours). Life experience (often a euphemism for the subject's age) was defined as the number of years between enrolling for the course and when the subject had graduated from high school. The three categories for this variable were: Less than three years, from three to five years, and greater than five years.

Secondary Analysis

Although these data were collected from 2000-2004, the first author agreed to analyze this data set as part of her doctoral dissertation in 2005 (*18*). Since these data had already been collected, the study represents an analysis of existing data, or a *secondary analysis*. According to Bryman (*20*), "secondary analysis is the analysis of data by researchers who will probably not have been involved in the collection of those data, for purposes that in all likelihood were not envisaged by those responsible for the data collection (p. 296)." The use of secondary analysis to analyze existing data sets has been advocated by educational researchers including the American Educational Research Association (*20–22*). Bryman (*20*) lists several advantages and disadvantages of analyzing other researchers' data. Advantages include the cost and time saved in using data that have often been collected over several years, that the new researcher's time can be devoted to data analysis, and that the secondary analysis allows new and different interpretations of the data set. Disadvantages include the fact that the new researcher has no control over how the data were collected and the quality of these data, and that the existing data set may be missing key variables or data that the new researcher would find valuable.

Results and Discussion

Table I contains the ANCOVA results for the 24-question posttest. Both covariates (pretest and GPA) were significantly associated with students' chemistry content knowledge. Once the ANCOVA has partialed out the variation attributed to these two covariates, the results for the independent variables in Table I were calculated. These results show that only course format was significantly associated with student content knowledge; comparison of the average scores for each group showed that students in the accelerated group learned more chemistry content (19.9/24) than students in the traditional course (17.7/24). Academic experience, life experience, and gender did not appear to affect the students' chemistry content knowledge.

The interpretation of these results are complicated by the fact that the 24 questions in the posttest are identical to the questions in the pretest. According to Borg and Gall (*23*), if a posttest is too similar to the administered pretest, students may show an improvement simply based on their exposure to the pretest. They also note that this effect tends to decline with increased time between the pretest and posttest. So, one may logically argue that since both groups of students took the pretest on the first day of classes, students in the 3- week accelerated course should be expected to perform better on the identical posttest than students in the 15-week traditional course.

Table I. ANCOVA Results with Posttest Scores as the Dependent Variable

Source	*df*	*Sum of Squares*	*Mean Square*	*F Ratio*	*p Value*
Covariates					
• GPA	1	653.67	653.67	85.87	0.000[a]
• Pretest	1	146.51	146.51	19.25	0.000[a]
Independent Variables					
• Course Format	1	212.49	212.49	27.92	0.000[a]
• Academic Experience	4	21.06	5.27	0.69	0.599
• Life Experience	2	6.27	3.14	0.41	0.663
• Gender	1	10.93	10.93	1.44	0.232
Error	207	1575.70	7.61		

[a] Significant association at $p \leq 0.050$.

In order to correct for this potential bias, we decided to run another ANCOVA analysis using the final examination scores, once the 24 pretest questions had been removed, as the dependent variable. This 76-point score still represents the instructors' view of the chemistry content knowledge these students should be expected to know after finishing this course, but since the students had never seen these questions, the familiarity bias mentioned for the posttest scores should not be applicable here. The results of this analysis appear in Table 2.

In this analysis, both of the covariates were found to be significantly associated with students' chemistry content knowledge as measured by the 76 remaining questions on the final examination. These results also show that course format and life experience (years past high school) significantly impact student chemistry content knowledge in these classes. Students in the accelerated course learned significantly more chemistry content (51.6/76) compared to the students in the traditional course (44.7/76). Students who graduated less than three years before enrolling in this course (46.5/76) and students who graduated from three to five years before enrolling in this course (47.0/76) demonstrated significantly lower chemistry content knowledge than students who graduated more than five years before enrolling in this class (51.0/76). Figure 1 contains a plot of these values, including error bars.

Table 2. ANCOVA Results with Final Examination Minus the Pretest as the Dependent Variable

Source	*df*	*Sum of Squares*	*Mean Square*	*F Ratio*	*p Value*
Covariates					
• GP	1	7913.77	7913.77	115.94	0.000[a]
• Pretest	1	1650.87	1650.87	24.19	0.000[a]
Independent Variables					
• Course Format	1	1979.23	1979.23	29.00	0.000[a]
• Academic Experience	4	237.65	59.41	0.87	0.483
• Life Experience	2	574.09	287.05	4.03	0.016[a]
• Gender	1	11.59	11.59	0.17	0.681
Error	207	14129.69	68.26		

[a] Significant association at $p \leq 0.050$.

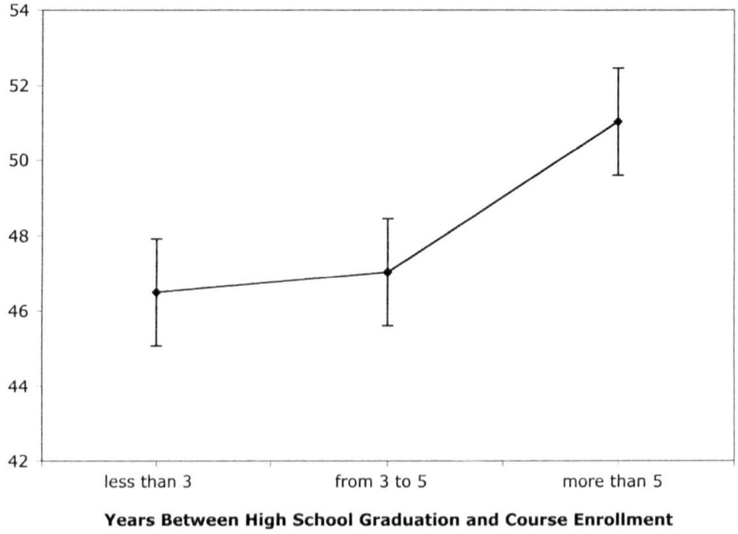

Figure 1. Plot of the significant main effect for the life experience variable.

Conclusions

Based on the results of this study, it appears that students enrolled in the accelerated 3-week summer versions of an introductory chemistry course demonstrated a more robust chemistry content knowledge than students enrolled in the 15-week fall versions of this course. This study also showed that students who had graduated high school more than five years before taking this course learned more chemistry content than students who had graduated five or fewer years before taking this course. A comparison of these results to those from a previous study (*19*) that looked at the final course grades for the same student sample shows that the differences identified in this study between students in the accelerated and traditional-length courses for their final examination scores persisted in the assignment of their final course grades. This study also identified differences in the final examination scores (after the pretest questions were removed) based on the time span between high school graduation and taking this course. However, these differences based on the life experience (age) of the students did not result in differences in the grade distributions ultimately assigned for these courses.

The superiority of accelerated over traditional-length classes in improving student success has been reported in several content areas (*1–3*, *5–7*, *10*, *11*, *13*, *15*, *16*); this is one of the first studies to look specifically at chemistry content. Greater student success in accelerated courses has been attributed to decreased course fragmentation, increased student-instructor interaction, increased student camaraderie, and greater student satisfaction (*1*, *2*, *6*, *8*, *10*, *11*). In this study, the improved performance of students in the accelerated course was attributed to the instructional pace and fewer academic distractions compared to the traditional chemistry course. The accelerated class covered the same material in the traditional course, but at a pace that was five times faster. This increased pace forced the students to focus on the material each day if they were to be successful in this course. The fact that this course met for five to six hours per day for five days a week, it was difficult for these students to take another course during the same semester, especially if they were working or raising a family. Taking only one class allowed these students to focus their attention on a single subject instead of splitting their efforts between two or more courses.

The literature results for life experience (age) are more limited, and somewhat mixed. Geltner and Logan (*7*) found that older students were more successful in traditional courses while Wlodkowski and Kasworm (*5*) and Caskey (*12*) reported that older students were more successful in accelerated courses. The results of this study showed that older students performed better on the final examination than their younger counterparts regardless of whether they were enrolled in the traditional or accelerated courses. We believe that the improved performance of the older, non-traditional students in this study can be attributed to the fact that these students tend to be more intrinsically motivated to succeed (*6*). Future research in this area should attempt to address the issue of student motivation and how it affects their performance in accelerated versus traditional chemistry classes. This research should also look at students' long-term retention of chemistry content knowledge to determine if the superiority of the accelerated class format persists over time.

This study was performed using existing data and showed several of the advantages and disadvantages of a secondary analysis listed by Bryman (*20*). Using this data set allowed for the analysis of four years' worth of data that could not have been easily collected as part of the first author's doctoral dissertation, and this allowed the researcher to focus on analyzing the data. These data were useful in answering the research question regarding student success and how it relates to course format, academic experience, life experience, and gender. However, had the researcher been involved in the data collection process, the data set would have been different. Having the pretest questions repeated verbatim in the posttest could have led to an unintentional bias based on the length of the course and students' memory of the questions. In addition, it might have been helpful to collect qualitative data in the form of student interviews to corroborate or refute the proposed explanations regarding the greater performance on the final examination for the older students and for those enrolled in the accelerated courses.

References

1. Hinger, B. *System* **2006**, *34*, 97–118.
2. Rikard, G. L.; Banville, D. *High Sch. J.* **2005**, *88*, 26–34.
3. Seamon, M. *Teachers Coll. Rec.* **2004**, *106*, 852–874.
4. Hsu, L. *Gifted Child Q.* **2003**, *97*, 212–218.
5. Wlodkowski, R. J.; Kasworm, C. E. Accelerated Learning for Adults: The Promise and Practice of Intensive Educational Formats. In *New Directions*

for Adult and Continuing Education; Number 97; Imel, S., Ed.; Jossey-Bass: San Francisco, 2003; pp 5–15.
6. Fall, L. T. *Journalism Mass Comm. Educ.* **2001**, *56*, 39–48.
7. Geltner, P.; Logan, R. The Influence of Term Length on Student Success. *Report for Santa Monica College*, April 20, 2001.
8. Logan, R.; Geltner, P. The Influence of Session Length on Student Success. *Report for Santa Monica College*, April 4, 2000.
9. Lee, M. M. Westchester Community College Summer School Survey, 1996 Session I and Session II. *Report for Westchester Community College*, Valhalla, NY, 1996.
10. Canady, R. L.; Rettig, M. D. *Educ. Leadership* **1995**, *16*, 4–10.
11. Fallon, K. Intensive education: How it affects teachers' and students' work conditions. In *Proceedings of the Annual Meeting of the American Educational Research Association, April 18–22, 1995*; San Francisco, 1995.
12. Caskey, S. H. J. *Cont. Higher Educ.* **1994**, *42*, 23–27.
13. Van Scyoc, L. J.; Gleason, J. *J. Econ. Educ.* **1993**, *24* (1), 15–22.
14. Williams, W. The relationship between class scheduling formats and the academic achievement of graduate students. In *Proceedings of the 2nd Annual Meeting of the Association of Teacher Educators, February 15–19, 1992*; Orlando, FL, 1992.
15. Scott, P. A.; Conrad, C. F. A Critique of Intensive Courses and an Agenda for Research. *Report for Wisconsin University*, Madison, WI, March, 1991.
16. Pflanzer, R. G.; East, J. R. *J. Coll. Sci. Teach.* **1984**, *14*, 110–114.
17. Kanun, C.; Ziebarth, E. W.; Abrahams, N. *J. Experiential Educ.* **1963**, *32*, 123–132.
18. Hall, M. V. D.A. Thesis, Middle Tennessee State University, Murfreesboro, TN, 2008.
19. Hall, M. V.; Wilson, L. A.; Sanger, M. J. *J. Chem. Educ.*, in press.
20. Bryman, A. *Social Research Methods*, 3rd ed.; Oxford: New York, 2008; pp 294–312.
21. American Educational Research Association: New Release for May 7, 2007. Media Contact: Helaine Patterson. For Researchers, Policymakers, and Funders American Educational Research Association (AERA) Report Provides Guidance for Establishing Causality through Both Experiment and Observational Designs. http://www.aera.net/uploadedFiles/News_Media/News_Releases/2007/NRCausalityRelease.pdf.
22. Duran, R. P.; Eisenhart, M. A.; Erickson, F. D.; Grant, C. A.; Green, J. L.; Hedges, L. V.; Levine, F. J.; Moss, P. A.; Pellegrino, J. W.; Schneider, B. L. *Educ. Res.* **2006**, *35* (6), 33–40.
23. Borg, W. R.; Gall, M. D. *Educational Research*, 4th ed.; Longman: New York, 1983; pp 824, 635.

Chapter 9

General Chemistry Student Surveys: Longitudinal Data about Which Factors Helped Students Learn

Norbert J. Pienta*

Department of Chemistry, University of Iowa, Iowa City, IA 52242
***E-mail: norbert-pienta@uiowa.edu**

A survey in which students assign value to course components that they believe helped them learn in a first semester general chemistry course was used over a ten year period to examine changes in administrative and curricular processes. Students were asked to judge the value of formal class meetings (i.e., lectures, discussions, and laboratories), course materials (i.e., books, study guides, end-of-chapter problems, and homework) and other potential interventions (i.e., friends, paid tutors, and study groups). These students also reported the amount of time spent on in-class and outside activities, including study time for the chemistry course. An analysis of the data shows that student use of many course materials has changed over the years but the choice of the textbook as one of the highest valued components remains relatively constant. One change that is evident from the data is that student ranking of study guides has decreased as the use of e-homework has become more prevalent.

Introduction

The first half of a two-semester general chemistry sequence is a gatekeeper course for many science, engineering, and pre-professional majors (*1*). Information about student activities and expectations in such a course could inform a broad range of curricular decisions. Starting in 1999 in anticipation of curricular changes and every few years for the next decade, students in a traditional two-semester general chemistry sequence were asked to rate course components and the impact of those items on their learning. The longitudinal nature of the study, based on large numbers of similar students, provides trends that define student behaviors, particularly those related to practices in the first semester of general chemistry. Thus, students were asked to judge the value of formal class meetings (i.e., lectures, discussions, and laboratories), course materials (i.e., books, study guides, end-of-chapter problems, and homework) and other potential interventions (i.e., friends, paid tutors, and study groups). In addition, the same students were asked to report the amount of time spent on in-class and outside activities, including study time for the chemistry course.

Background

University of Iowa students (ca. 30,000 total of which about 22,000 are undergraduates) typically originate from Iowa (60%) or surrounding states (25%). Undergraduate students come from the top half of their high school class. The University's College of Liberal Arts and Sciences, where a large majority of students first enroll, requires three mathematics and three science courses as part of the undergraduate curriculum. In anticipation of course pre-registration during summer orientation, students take mathematics and chemistry diagnostic tests online. Based on their scores, about 900 students place into a one-semester preparatory chemistry course and another 900 into the first semester of the two-semester sequence, Principles of Chemistry I. The "Principles" course is offered as three lecture sections (team taught by three instructors) that meet 3 hours per week; a discussion section (24 students per section that are directed by teaching assistants for one hour per week); and an integrated laboratory component. The laboratory portion comprises two parts: a "case study session" directed by an instructor that meets for 1.5 hours in alternating weeks with the traditional laboratory experiment section, which meets for 3 hours the week following the corresponding case study session. The case study session involves presentation of material, work on activities, and discussions in anticipation of the experimental work the following week. This laboratory design was implemented in the fall of 2002. Students in prior years enrolled in the laboratory as a separate 2 credit course. More details about the course redesign and its assessment have been reported (*2*).

A survey was created to determine the perceived value of course components in Principles I (*3*). The paper instrument asked students to assign a value on a Likert scale (1 = lowest, 10 = highest, 0 or blank = not utilized) to each item listed in Table I. The survey was not administered until early in the spring semester so that the students were describing the entire experience of the Principles I course completed in the previous fall. The intention was to collect data about the course in the absence of anxiety related to examinations or course requirements, and after students had had some time to reflect on the experience. Thus, the data reported herein are limited to those students who were enrolled in the second semester, Principles II. This eliminated students who withdrew or failed Principles I or who only chose to take one semester of chemistry. Most majors that require the course call for both courses of the sequence. Engineering majors are often an exception because most types of engineering only require Principles I. In spring 2003, the survey was administered to the engineering students who had taken Principles I the previous fall by collecting data in an introductory engineering methods course in which all were enrolled.

Survey Components and Data

Table I contains the numerical data collected between 1999 and 2009 in answer to the question, "What helped you learn and succeed in Principles I?" Each student was asked to enter a number between 1 and 10 for each item and was instructed to leave an item blank or enter a zero if an item was not relevant to them. The components that were evaluated are listed in the first column. Values appear for several different years in the subsequent columns. The years in the table were the only ones in which the data were collected. The first number in columns 2-6 is the average of the student scores, while the percentage in parentheses represents the portion of respondents that evaluated that item. For example, the average rating for the textbook in 1999 was 7.7 and 100% of the 261 surveys had a value entered.

The first issue is the matter of sampling. For all years except the last (i.e., 2009), paper surveys were administered and collected in lecture classrooms early in the semester. In 2009, paper surveys were collected in discussion sections. Table II shows the number of surveys compared to the students enrolled in the spring course where the surveys were administered and the enrollment of the previous fall for which the questions were answered. For the enrollees, both the number and the percentage (in parentheses) are reported. Thus, in the second column with the "Fall 1999" heading, there were 261 surveys collected out of a spring enrollment of 423 students (in Principles II) regarding the fall course (Principles I) for which the enrollment was 712. The percentage of respondents to enrollees is 62% and 37%, respectively. For the fall 2002 class, additional data were collected from the engineering students using an engineering class they had in common to administer the surveys. Those data are very similar the "Fall 2002" column in Table I. It is not included here because it was not collected in any other years.

In order to gauge how representative the survey values are, a population analysis is appropriate. The attrition in enrollment between the fall offering of Principles I and the spring offering of Principles II can be accounted for by the number of engineers for whom the second course is not required; the students that did not succeed in Principles I; and others who changed majors or left the University. For example, the rate of D, F and W grades in the Principles I course has been at ca. 15% during this period (i.e., about 100-150 students). Chemical and biomedical engineers enroll in both semesters of general chemistry, as do other students who are also classified as pre-medicine. An estimated 120-160 engineering students take one chemistry course, and based on 2002 data, their data appear to match that of the remaining class population. As a result, the survey is representative of the students from the fall who succeeded in that course (i.e., achieved a grade of C- or higher). Furthermore, survey sampling is consistent across the years, making year-to-year comparisons appropriate.

Table I. Student Survey Data for "What helped you learn and succeed…"

1= least to 10 = most	*Fall 1999 N = 261*	*Fall 2000 N = 228*	*Fall 2002 N = 204*	*Fall 2005 N = 228*	*Fall 2009 N = 460*
textbook	7.7 (100%)	8.1 (100%)	7.7 (100%)	7.3 (99%)	7.2 (99%)
solutions/study guide	6.8 (79%)	7.4 (90%)	7.0 (93%)	6.6 (75%)	4.5 (65%)
CD-ROM w/ text	4.0 (68%)	2.5 (56%)	2.4 (56%)	1.6 (34%)	---
text website	---	---	2.7 (55%)	2.5 (40%)	3.3 (43%)
lecture content	5.6 (99%)	6.5 (99%)	6.6 (98%)	7.4 (98%)	6.7 (96%)
lecturer interesting	4.5 (95%)	5.0 (97%)	6.4 (98%)	6.5 (95%)	6.0 (94%)
lecturer relevant	5.7 (96%)	6.3 (96%)	6.6 (99%)	7.0 (96%)	6.5 (95%)
course website	---	---	6.5 (94%)	5.9 (89%)	6.3 (89%)
EOC problems	---	---	7.1 (94%)	6.4 (91%)	5.7 (73%)
written homework	6.6 (98%)	5.4 (72%)	---	---	---
electronic homework	4.1 (42%)	3.6 (76%)	7.6 (98%)	7.0 (96%)	7.5 (98%)
discussion session	5.9 (97%)	4.3 (87%)	5.2 (98%)	6.0 (95%)	6.8 (98%)
case study session	---	---	4.9 (98%)	5.8 (96%)	6.4 (97%)
laboratory session	---	---	4.5 (98%)	5.3 (95%)	5.9 (96%)
TA office hours	4.1 (51%)	3.4 (52%)	3.0 (33%)	4.8 (46%)	5.4 (52%)
study group	5.9 (49%)	4.3 (46%)	4.7 (58%)	5.0 (45%)	6.0 (53%)

1= least to 10 = most	***Fall 1999*** *N = 261*	***Fall 2000*** *N = 228*	***Fall 2002*** *N = 204*	***Fall 2005*** *N = 228*	***Fall 2009*** *N = 460*
paid tutor	3.5 (18%)	2.4 (26%)	1.9 (24%)	4.3 (18%)	2.7 (23%)
friends	6.8 (83%)	6.3 (73%)	6.1 (69%)	7.0 (80%)	6.8 (78%)

Table II. Number of Surveys and Percentage of Class Represented

	Fall 1999	*Fall 2000*	*Fall 2002*	*Fall 2005*	*Fall 2009*
number surveys	261	228	204	228	460
enrollees, spring course	423 (62%)	354 (64%)	468 (44%)	492 (46%)	558 (82%)
enrollees, target fall course	712 (37%)	677 (34%)	814 (25%)	850 (27%)	895 (51%)

Statistical analyses or comparison of data between years warrant consideration of frequency distributions of scores for individual items. Plots of the distribution of Likert scale scores versus the frequency of each score for fall 2009 data for textbooks, solutions manuals (a.k.a., study guides) and lecture content appear in Figure 1. The frequency scales are identical for all three cases. The textbook data approaches a single-tailed distribution, while the lecture content values are more evenly distributed. The solutions manual/study guide values appear bimodal with one set of values centered between 1 to 2 and the other between 6 and 7. Other examples (not shown) include ones with low average values like "paid tutors" or the text CD-ROM for which the frequencies are broadly distributed. The frequency distributions for textbook and lecture content are high, leading to the single-tail pattern. In the case of the study guide, it appears that the distribution is bimodal with a peak at 1.5 and 6.5. A possible explanation is that the lower-score group finds very little value to these manuals.

Over the course of ten years, the instructors in a course change regularly as do the textbook and accompanying course materials. In fact, comparison of instructors from one year to another is confounded by having multiple instructors in the course. Depending on the actual enrollment, either 2 or 3 lecture sections would be used each fall, generally with one faculty instructor per section. In fact, there is some homogenization because the course is team taught—a single instructor teaches all sections for some portion of the course. The intention is to equalize the experience. Common evening examinations for the entire cohort are another course feature. As a result, the reported ability of these instructors to make the lecture or presentation interesting and relevant represents an integrated evaluation. (It also enables the presentation of the data here without singling out individual instructors.) Thus, each of the columns in Table I represents a unique set or combination of lecture instructors. The instructor for each section of record appears as an entry in Table III; for example, one instructor taught both sections in fall 2002.

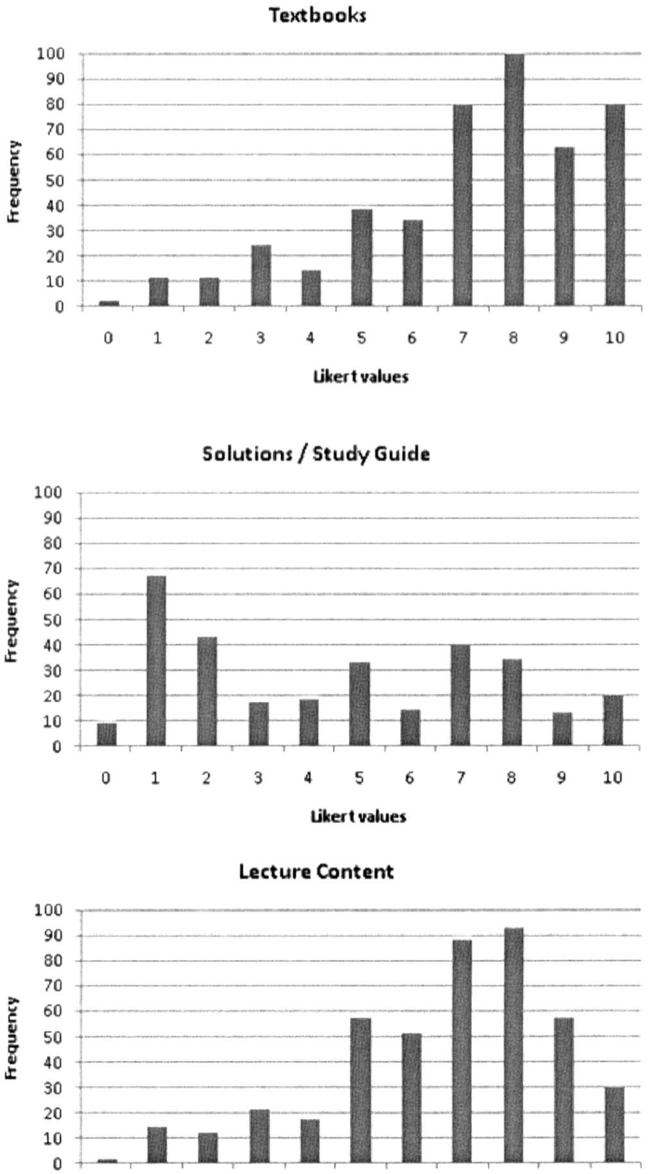

Figure 1. Three Plots of Survey Scores vs. Frequency for Fall 2009: Textbook, Solutions Manual/Study Guide, and Lecture Content.

Table III. Unique Instructors, Textbooks and e-Homework Products

	Fall 1999	*Fall 2000*	*Fall 2002*	*Fall 2005*	*Fall 2009*
Instructors*	A, B	B, C	C, C	B, D, E	D, D, F
Textbooks**	A	A	B1	C	B2
e-Homework***	A	B	C	D	D

* Letters represents different instructors. ** Letters represents different textbooks. Numbers represent different editions of the same textbook. *** Letters represents different iterations of e-homework.

The textbook was also changed several times over the decade (see Table III). Textbook A was used in 1999 and 2000. Textbook B1 was used in 2002, and a different edition of it (B2) in 2009. Textbook C was used in 2005. Given the 3-year lifetime of a text edition, the changes are not unusually frequent. All three texts can be characterized as common, top-5 selections. In the years including 2002 and subsequent to it, commercial electronic homework was selected in tandem with the textbook packages. In 1999, electronic homework (e-Homework A) was used on a voluntary basis—it was available as a supplementary activity without course credit. In 2000, e-Homework B was required; the user interface and data entry were not very "friendly" and the distribution of problems was skewed toward difficult ones. Another product that offered tutorial help and feedback was adopted in 2002 (e-Homework C) and yet another (e-Homework D) has been used for the data in the last two columns in Table I. Electronic homework has been required since 2002, and its value represents about 10% of the course grade. A summary of e-Homework products also appears in Table III. In 1999 and 2000, written homework was assigned from the end of chapters problems (i.e., EOC problems), collected, and graded. Teaching assistant resources were limited and as a result only a small percentage (ca. 10%) of randomly selected problems from the written homework were actually graded.

Analysis and Discussion of Results

The data in Table I can be considered in two ways: the absolute or relative score of each component or the changes of the scores or ranks over time. In several instances, the latter are correlated with administrative or procedural changes within the course. In order to compare the various course components, the Likert scale scores were converted to percent rank scores, a conversion that makes year-to-year comparisons clearer. The minimum absolute score becomes the zero percent rank score, the maximum absolute score becomes the 100 percent rank value, and intermediate values are scaled accordingly. Plots of the percent rank scores for each year and each item appear in Figures 2 and 3.

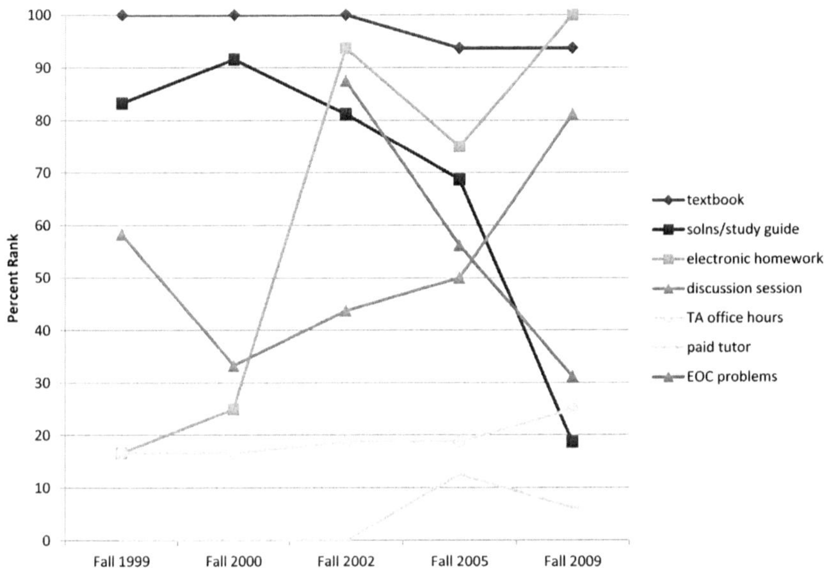

Figure 2. Percent Rank Score by Year for Selected Course Items.

For all but the last year (i.e., fall 2009) the textbook represents the highest ranking item or is tied for that position. This is remarkable given the popular belief that students no longer buy or use textbooks. However, discussions with the managers of the two major bookstores in town confirm that Iowa students buy (and sell) chemistry textbooks for the Principles courses at sufficiently high numbers to corroborate the ranking. (Students buy and sell books to each other and on the Internet so a detailed analysis was not attempted.) The high cost of books is an issue on this campus, and the topic has been debated in the student press (*4*, *5*). There is no evidence to suggest that Iowa students regard books more highly than other students; national survey data about student use of textbooks has been collected by the TextRev website (now inactive) and the results have been previously reported by Smith and Jacobs (*6*). How students use those textbooks listed on the site may not match instructors' expectations and may have changed since the report was published. This is supported by the decline in rank for the textbook between 2005 and 2009.

On the other hand, the rank of the solutions manual or study guide that accompanies the textbook has changed markedly since 2000. This decrease provides some insight into student study habits and is the result of changes in course practices. During the same time, the perceived value of traditional written homework and end-of-chapter (EOC) questions has also decreased. Students currently get credit for the successful completion of electronic homework and no tangible rewards for completing the EOC problems that are assigned but not graded or otherwise evaluated. The decline in rank of the study guide and the EOC questions are remarkably similar, again pointing to the historical use of the solutions manual. Students depended on the study guide to help them with the assigned problems. Even though the electronic homework may have problems taken directly from the EOC questions, the study guide is not perceived in the same way by students. In fact, calculated problems on the electronic homework often are set up to allow 3-4 attempts and do provide some help and tutorials. When confronted with only one or two attempts remaining, many students seek help from instructors and teaching assistants, as well as friends. But the written explanation that appears in a study guide has lost considerable favor in this regard.

In order to promote the use of discussion sections and the teaching assistants, the instructors in recent years have offered a small amount of credit (ca. 5%) for attending and participating in teaching assistant-led discussion sections. As a result, the perceived value of discussion sections has increased since 2002. Students go to the discussion section to get the credit, and many additionally find something of value. It is not clear whether the "free" points are the extent of this value. If the students had learned to value their teaching assistants, one might expect an increase in the value assigned toTA office hours. TAs are required to spend 2 hours per week of "office hours" in a special classroom, and those collective hours are spread out over the entire week to maintain a resource center that is manned 40-50 hours per week. If the value of the discussion was simply the TA, quality time with a smaller student-to-TA ratio during office hours might be more highly appreciated by students. This is not to say that the individual TAs are not valued—there could be several reasons why discussion sections and TA office hours don't track each other (e.g., student time constraints or inability to overlap with one's TA's hours).

Paid tutors consistently rank at the bottom of the scale for interventions. The Chemistry Department has maintained lists of potential tutors but little or no effort goes into evaluating their ability to help or assessing their success at doing so. Multiple venues exist for peer tutoring, including some students that are recruited and trained by Residence Life staff to work evenings in various dormitories, members of a professional service organization (Alpha Chi Sigma), and other programs (Women in Science and Engineering, support groups for disadvantaged students, and the tutoring program provided by the Athletics Department). Peer tutors in a Supplemental Instruction program were available until the course redesign in 2002, but the program was phased out when student participation dwindled. Supplemental Instruction SI was successful historically (*7*, *8*), as have been more recent versions like PLTL (*9–12*).

In part, the student survey was devised to look at some changes in the Principles sequence before and after a curricular change. Some items that were introduced or terminated appear in Figure 2. The course redesign was implemented in 2002, and was accompanied by course components that were either added or deleted. The course redesign involved the integration of a separate laboratory course into both semesters. Thus, in the old system students took two 3-credit-hour lecture courses (with three 50-minute lectures/presentations and one 50-minute discussion section) and one 2 credit hour laboratory course; the second semester lecture course was the co- or pre-requisite for the lab. In the new system, the laboratory is fully integrated with the lectures and discussion, creating two 4-credit-hour courses. The laboratory involves two distinct venues—the so-called case study session and the actual laboratory. The former is convened for 90 minutes and serves as the preparation for the experiment in the following week.

The case-study section takes its name from a contextual scenario and the meeting time is filled with activities, discussions, safety and technique demonstrations and some didactic instruction.

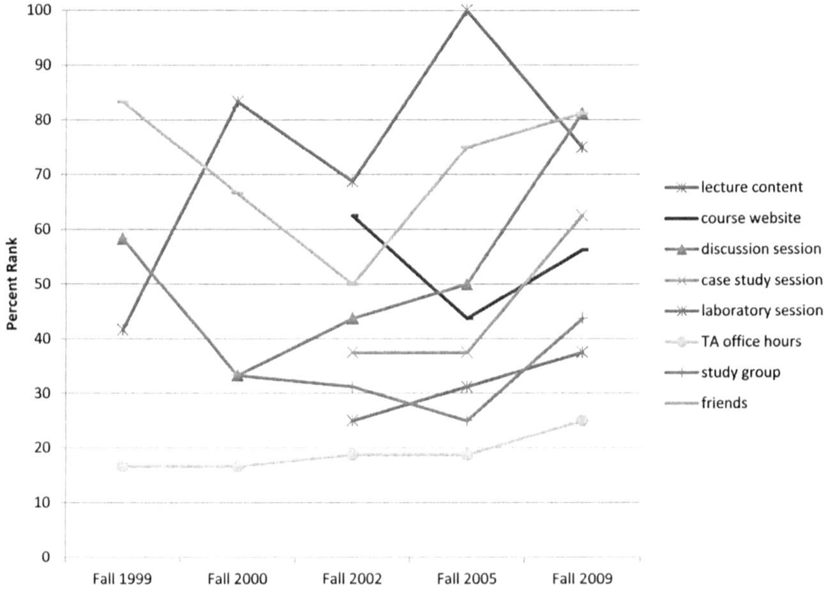

Figure 3. Percent Rank Score by Year for Additional Course Items.

For example, the first experiment is a qualitative analysis lab. In the case study session, among other things, the students must devise a flow chart to successfully identify an unknown from a three-component system. The process prepares them for the following week when they must solve the identity of an unknown from a group of twelve. For the energy and thermodynamics experiment, the context is the energy balance for a runner—given the exercise routine and the amount of calories consumed at meals, does the athlete need to eat a power bar each day? The experimental session, which meets for 3 hours, involves the calorimetric determination of energy contained in food items. Each case study session includes all sections of students whose experimental lab would meet at the same time (typically 2-4 sections or 48 to 96 students). The case study session is directed by an instructor rather than teaching assistants. The experimental sections, which always occur the week following case study sessions, are conducted by teaching assistants.

The guided inquiry in the two different laboratory sessions in chemistry's redesign does not have a parallel in other introductory science courses on campus. Students appear to have a better grasp of expectations and how to succeed using the guided inquiry laboratories compared to traditional laboratories. Both case study and laboratory sessions appear in the data in Figure 3, and both have increased in their rankings slowly since their introduction in 2002 as our experience with them improved. Homework written on paper and graded by teaching assistants was phased out as an efficiency step and served to provide the additional TAs that became necessary after the course redesign. The publisher's website and a CD-ROM accompanying the book were phased out as unpopular (and perhaps unnecessary). The latter two items (TA graded homework and publisher's website/CD-ROM) never appeared to be very useful in the opinions of the students surveyed.

Lecture content data is listed in Figure 3 and appears well regarded, although the values are not consistent. Perhaps the relatively high rank for this feature is not surprising because the "lecture presentation" is a substantial portion of the students' time commitment in the course. In fact, most students do not have a good appreciation for how they actually learn (*vide infra*). Another likely reason for the high scores for lecture content comes from the population that was surveyed. Because the survey data were collected in class, those who attend lecture value this course component—they already declared their opinion by showing up. In fall 2009, a high percentage of students were polled (Table II, 82% of the Principles II enrollment) and the perceived value of lecture is still quite high. Another survey item is the course website, accessible via course management systems like WebCT and Desire2Learn. Most instructors have provided outlines of the course lecture notes, practice examinations, and other tutorial materials on the course website. It is not clear whether the lecture content and/or the course website refer to the instructor's lecture notes item on the rankings.

Several of the instructors encouraged the formation of study groups outside of class, particularly during the earlier years reported in this study. Those groups were intended to promote discussions and to help develop constructive habits of mind. In addition, especially during the early years, students reported their friends as valuable to success in the course, something they also reported during focus groups. Study groups were formed using fellow students that lived nearby in a residence hall or other students they met in discussion or lab. Friends were people they knew (and presumably trusted) from high school. Unfortunately, no organized attempts were made to exploit the apparent useful relationship between students and their friends.

Accounting for Student Use of Their Time

Students were also asked to report how they spent their time in several categories as part of the course component evaluation. Table IV contains a summary of the reported time in class and outside of class, including time on task working on chemistry. Because it seemed useful to be able to compare this with the other activities in which the students would be engaged, several additional categories of out-of-class activities were examined. The standard deviations in Table IV are quite large, implying considerable variability among the respondents but also suggesting that the students at this stage in their academic careers are not very skilled at estimating their time. The average values for the in-class time commitment are quite reasonable and consistent with schedules for science majors. A typical 15 credit hour course load might include a laboratory or two, resulting in average numbers like those reported here. The out-of-class values are more variable, resulting in higher standard deviations. The average values are quite disappointing to many instructors because they are not the 2-3 hours outside class for each hour in class. The times reported for studying chemistry are consistent with the proportion of hours compared to a full load. A 4-credit-hour Principles course is slightly more than 25% of the total course load and the 5-7 hours studying chemistry is consistent.

A substantial but not a great percentage of Iowa students work for wages, and the number of hours per week is not overwhelming. On the questionnaire, several examples of "living" were listed to clarify the category (e.g., eating, sleeping, personal hygiene, laundry, moving among activities). A week comprises 168 hours and some students accounted for half or less of their time. The survey form did provide the number of hours in a week; for example, the total of the averages in the column is 150 and 156 hours for 1999 and 2009, respectively. New categories like sending text messages or viewing Facebook® pages may be appropriate for future surveys.

Table IV. Students' Reported Use of Their Time

Weekly Schedule (avg hrs)	*Fall 2000, hrs*	*Fall 2002, hrs*	*Fall 2005, hrs*	*Fall 2009, hrs*
in class (all courses)	16 ± 4	18 ± 5	18 ± 4	19 ± 7
studying out of class (all courses)	19 ± 11	23 ± 15	20 ± 10	23 ± 15
(gen chem)	5.6 ± 4.1	5.0 ± 3.2	5.0 ± 3.0	7.8 ± 6.2
(e- homework)	4.0 ± 4.2	2.9 ± 2.0	3.0 ± 1.5	4.4 ± 4.2
working at a job	14 ± 7 (35%)	15 ± 8 (30%)	15 ± 10 (28%)	5 ± 11 (32%)
entertainment	14 ± 10	22 ± 18	17 ± 11	21 ± 18
"living"	77 ± 30	70 ± 27	77 ± 23	76 ± 36

Conclusions

Although not framed by a theory, surveys like the one reported here have been used to inform decisions about course curricula and pedagogies (*13*). Iowa faculty used these data to inform changes made during the implementation of the course redesign; the implementation of electronic homework; changing homework from "paper" to electronic; awarding small point value to discussion sections in order to improve attendance; and other procedural changes. A distinction is made here between what actually helps students learn versus what students value among course components. The latter informs educators about where to look for answers while the former provides those answers. One must know what the behaviors actually are before attempting to explain them. For example, one can speculate about whether textbooks are the best tools for student learning or even generate a proper research study to understand the issue. The data here show that students have apparently not abandoned the textbook and seem to find it valuable.

Seasoned instructors who interact with large numbers of students in these classes can report the myriad of ways that a textbook can be used—as a reference; to clarify what an instructor said; to find algorithms to solve word problems; to find data and information to use in solving problems; or a means to attain some level of conceptual understanding of the content. A research study is necessary to demonstrate the efficacy of the course component or to elucidate which reason(s) motivate the use that makes the difference.

The student surveys and their administration over a long period of time allowed the instructors in the general chemistry program at the University of Iowa to understand and respond to student behavior in the Principles courses. In some cases, the outcomes are correlated with student survey data, and in other instances when additional information was available from focus groups and other sources, more certainty can be ascribed to the changes. Many of the observations are generalizable to other programs and institutions. Some are worthy of extensive and formal explanation using the techniques and methodologies contained in this volume and others that have appeared previously (*14*, *15*).

Acknowledgments

The author acknowledges the efforts of Russell Larsen, Lynne Cannon, Sonja Franklin, Jason Telford, and Dwight Tardy in devising and implementing various components of the Principles of Chemistry redesign. The cooperation of fellow instructors Ed Gillan, Peter Hansen, Darrell Eyman, Sonja Franklin, Russell Larsen, Amy Michel Strathman, John Krstinansky, and Mohamed El-Maazawi is also recognized.

References

1. Tai, R. H.; Sadler, P. M.; Loehr, J. F. Factors influencing success in introductory college chemistry. *J. Res. Sci. Teach.* **2005**, *42* (9), 987–1012.
2. Pienta, N. J. From Course Redesign to Curricular Review: Assessment in Chemistry at the University of Iowa. In *Assessment in Chemistry Programs: Evolving Best Practices*; Ryan, J., Muffo, J. , Eds.; Assessment in the Disciplines Series; Association for Institutional Research (AIR): Tallahassee, FL, 2010.
3. Pienta, N. J. Innovations in Lecture: Some Cases of Student and Faculty Response. In *College Chemistry Teacher Survival Guide*; Bunce, D., Muzzi, C., Eds.; Prentice-Hall: New York, 2003; pp 85–94.
4. Messer, J. New Textbook Law May Ease Cost. *The Daily Iowan*, June 23, 2010. http://uwire.com/2010/06/23/new-textbook-law-may-ease-cost/.
5. Wahls, Z. Professors, Feel Our Pain. Opinions, *The Daily Iowan*, June 25, 2010. http://www.dailyiowan.com/2010/06/25/Opinions/17686.html.
6. Smith, B. D.; Jacobs, D. C. A window into how general and organic chemistry students use textbook resources. *J. Chem. Educ.* **2003**, *80*(1), 99–102, DOI: 10.1021/ed080p99.

7. Webster, T. J.; Hooper, L. Supplemental instruction for introductory chemistry courses: A preliminary investigation. *J. Chem. Educ.* **1998**, *75* (3), 328–331, DOI: 10.1021/ed075p328.
8. Mason, D.; Verdel, E. Gateway to success for at-risk students in a large-group introductory chemistry class. *J. Chem. Educ.* **2001**, *78* (2), 252–255, DOI: 10.1021/ed078p252.
9. Tien, L. T.; Roth, V.; Kampmeier, J. A. A course to prepare peer leaders to implement a student-assisted learning method. *J. Chem. Educ.* **2004**, *81* (9), 1313–1321, DOI: 10.1021/ed081p1313.
10. Gafney, L.; Varma-Nelson, P. Evaluating peer-led team learning: A study of long-term effects on former workshop peer leaders. *J. Chem. Educ.* **2007**, *84* (3), 535–539, DOI: 10.1021/ed084p535.
11. Hockings, S. C.; DeAngelis, K. J.; Frey, R. F. Peer-led team learning in general chemistry: Implementation and evaluation. *J. Chem. Educ.* **2008**, *85* (7), 990–996, DOI: 10.1021/ed085p990.
12. Gosser, D. K., Jr.; Kampmeier, J. A.; Varma-Nelson, P. Peer-led team learning: 2008 James Flack Norris Award Address. *J. Chem. Educ.* **2010**, *87* (4), 374–380, DOI: 10.1021/ed800132w.
13. Scantlebury, K.; Boone, W. J. Designing Tests and Surveys for Chemical Education Research. In *Nuts and Bolts of Chemical Education Research*; Bunce, D. M., Cole, R. S., Eds.; ACS Symposium Series 976; American Chemical Society: Washington, DC, 2008; Chapter 10.
14. Bodner, G. M.; Orgill, M. *Theoretical Frameworks for Research in Chemistry/Science Education*; Pearson/Prentice Hall Publishing: Upper Saddle River, NJ, 2007.
15. Bunce, D. M., Cole, R. S., Eds.; ACS Symposium Series 976; American Chemical Society: Washington, DC, 2008.

Chapter 10

Using Item Response Theory To Identify and Address Difficult Topics in General Chemistry

Kimberly D. Schurmeier,[a] Carrie G. Shepler,[b] Gary J. Lautenschlager,[c] and Charles H. Atwood*,[d]

[a]Chemistry Department, Box 8064, Georgia Southern University, Statesboro, GA 30458
[b]School of Chemistry and Biochemistry, Georgia Institute of Technology, 901 Atlantic Drive, Atlanta, GA 30332-0400
[c]Psychology Department, University of Georgia, Athens, GA 30602
[d]Chemistry Department, University of Utah, Salt Lake City, UT 84112
*E-mail: chatwood@chem.utah.edu

We have employed the modern psychometric analysis tool Item Response Theory (IRT) to analyze the results of 10 years of computerized testing and final exam results on over 20,000 students. Our analysis indicates that there are 8 especially difficult general chemistry topics which consistently cause problems for our students. These topics are: 1) the Particulate Nature of Matter, 2) Molecular Polarity and Intermolecular Forces, 3) Understanding Quantum Numbers, 4) Use of the Terms Strong, Weak, Concentrated, and Dilute in Relation to Acidic or Basic Solutions, 5) Molecular Image Problems, 6) The Mole Concept, 7) Solution Calorimetry, and 8) Inorganic Nomenclature. Once identified, we hope that educators will emphasize these topics in their teaching yielding improved success rates in their classes.

Introduction

Many students see general chemistry as a difficult course, and they often regard this course as an enormous obstacle to their academic and career goals. Some students believe that general chemistry is designed to "weed out the weak" (*1*). Chemical educators realize that in reality, general chemistry simply is a course that requires high level conceptual understanding in addition to algorithmic proficiency, a combination of skills that many students must work to develop. This study's primary goal was to identify the topics that are most difficult for students to learn so that instructors could better facilitate students' learning of chemical concepts and development of the skills necessary to succeed in general chemistry. A second research goal was to determine whether the difficulty of the course was consistent from year to year.

To accomplish these goals, examinations given to students over a four-year span from 2005 to 2009 were analyzed using Item Response Theory (IRT) and questions that did not thoroughly assess students' understanding of a topic were removed from the question pool for future examinations. Questions were then grouped by topic to determine those topics most challenging for students.

In this chapter we shall introduce the basics of IRT then show how we have employed IRT to determine those topics which are most difficult for our students as well as those that are especially easy. We hope that this will spur educators to pay special attention to the difficult topics. We shall also allude to our future plans to use this knowledge to identify "at risk" chemistry students early in the teaching process.

Item Response Theory

Over the last thirty years, IRT has been employed to analyze students' abilities, or success rates, in a wide range of subjects (*2–8*). Compared to Classical Test Theory (CTT), IRT provides more useful detail on assessments. In particular, by analyzing patterns in the assigned IRT question difficulties, chemical topics that are especially difficult for students can be determined. Most chemical educators are familiar with CTT, and here the major differences between IRT and CTT are highlighted to provide a better understanding of the practicality of IRT.

Both methods provide insight into certain aspects of assessment. CTT can be employed for small sample sizes and is simpler to perform. IRT is statistically valid for larger sample sizes, preferably 200 or more. Analysis with IRT involves choosing between one of the three possible IRT models, manually adjusting the model parameters, and then allowing a computer program to iterate through the data until the model converges. Using CTT, the mean, median, and Gaussian probability distribution for the test can be calculated. For each test item, CTT can also determine an item discrimination factor, by comparing the performance of the top quartile of students versus the bottom quartile on that question. However, CTT depends more on the subject group whose exams are being analyzed, as well as upon the nature of the examination, than IRT. For example, if the same group of individuals were given two different assessments on the same subject using different assessment items, CTT analysis would likely yield different results for the two assessments. When model assumptions prove reasonable, IRT analysis is independent of the individuals assessed and the assessment items used. Rather than assigning a mean and median, IRT assigns each student an "ability level" based upon their responses to the assessment items. To a first approximation, a student's IRT ability level is a measure of their proficiency on the material being assessed. Additionally, each item on the examination is assigned a corresponding difficulty level based upon those students with a given ability level (and higher) who correctly answered the question. After the IRT analysis is performed, each test item's difficulty indicates how that item discriminates between students within the entire ability range. This is an important distinction between CTT and IRT that we have extensively employed.

Assessment reliability can be determined with both CTT and IRT. In CTT the examination reliability is determined by calculating Kuder-Richardson 21 (KR-21) values, the value of which is measured relative to the test mean. With IRT, examination reliability is found by generating a total information curve that determines how accurate the assignment of student abilities is across the entire ability scale. The examination reliability described in this chapter will be measured by calculating both the KR-21 values and total information curves (*9*, *10*).

Although many explanations of the fundamentals of IRT can be found in the literature (*9*, *11*, *12*), we provide the basic mathematical details here for clarity. The IRT program Bilog MG3 was used for the data analysis fitting procedure. IRT analysis constructs an item characteristic curve (ICC) for each assessment item using three-parameters in Equation 1; where b is the item difficulty parameter, a is the item discrimination parameter, c is the pseudo-guessing parameter and θ is the person ability level. Item difficulty and person ability are measured on a common metric referred to as the ability scale. The ICC shows a plot of the probability, $P(\theta)$, that a student with a person ability level, θ, will correctly answer the question being analyzed.

There are three possible IRT models. In each case, the basic IRT equation is employed, as shown in Equation 1. Student response data, in the form of a file containing 0's for incorrect responses and 1's for correct responses to each question, is iteratively fit to the IRT equation.

$$P(\theta) = c + (1-c)\frac{1}{1+e^{-a(\theta-b)}} \qquad (1)$$

The number of parameters used in each of the three models is variable. The one-parameter model, known as the Rasch model, employs only the difficultly parameter, b, which describes how difficult (or hard) a question (or item) is. Bilog MG3 displays the b value on a scale of typically -4 to +4 where -4 is an especially easy question and +4 is a particularly difficult item. However, some especially difficult items may exceed +4. The two-parameter IRT model employs the difficulty parameter, b, and the discrimination parameter, a, which describes how discriminating a question is between students with different personal abilities. Personal ability levels for each student are determined from the IRT parameter estimates (IPEs). Once IPEs are obtained and stored, Bilog MG3 is applied a second time to the data to generate estimates of a personal ability value for each examinee using either a maximum likelihood or a Bayesian estimation approach. For a non-discriminating question, IRT analysis yields an a parameter of zero, indicating that every student can answer the item correctly. Discriminating ability of the question increases with *a*, where values of approximately 1 or higher are considered to be discriminating questions. The most complex IRT model, the three-parameter model, incorporates the aforementioned parameters, augmented with a pseudo guessing parameter, *c*, designed to indicate the probability with which a student can "guess" the answer correctly. Theoretically, a multiple choice question with four possible answers has a *c* value ≈ 0.25. However, if distracters in multiple choice questions are not carefully designed, the pseudo guessing parameter can have values greater than 0.25. In theory, the pseudo guessing parameter for non-multiple choice questions should be close to zero. de Ayala (*9*) as well as Embretson and Reise (11) describe the mathematical details of how IRT determines the pseudo guessing parameter. Essentially Bilog MG3 attempts to estimate the value of the lower asymptote of the item response curve based upon the responses of lower ability students.

The one-parameter Rasch model was not used in our analysis because the inherent lack of flexibility proved too restrictive (*13–15*). All of the discussed examinations were analyzed using either the two- or three-parameter item response theory logistic models. As noted previously, a multiple choice question will have an associated guessing factor due to the limited number of responses, while there are, in principle, an infinite number of responses for a free-response question. Therefore, the two-parameter model is optimal for free-response question analysis, while the three-parameter model is most suitable for the analysis of multiple choice questions. Certain free response questions did not fit the two-parameter model well. In these cases, the three-parameter model was employed. This approach of adjusting the analysis to suit the problem affords a more rigorous analysis than the approach taken in previous studies in which a single model was assumed and questions not fitting that model were excluded from the analysis (*3*, *5*, *6*, *8*). Even with the more complex three-parameter model being used for many questions, there were still items that did not fit the model properly.

Ideal item characteristic curves are sigmoidal with a steep slope (greater than 1), indicative of the *a* value, and a midpoint that represents the *b* value. An ICC with a slope equal to zero indicates there is an equal probability of every student answering the question correctly regardless of that student's personal ability. In the limit of an infinite slope, the ICC will be a step function centered at an ability level *b*. In this case, a student with an ability level greater than *b* will always get the correct answer, while a student below this level will only answer the question correctly by guessing. An example of a highly discriminating ICC is shown in Figure 1.

The slope, *a*, of the curve in Figure 1 is 2.566 indicating a highly discriminating question. The item ability, *b*, determined from the item's ICC is 1.074 indicating that students with a person ability of $\theta = 1.074$ have a probability of 0.533, calculated using Equation 2, of correctly answering the question (*12*). Students with ability less than 1.074 have a decreasing probability of correctly answering the question whereas students with abilities greater than 1.074 have an increasing probability of correctly answering the question. Note that the guessing parameter, *c*, for this question (indicated by the lower asymptote) is 0.066. This indicates that all students, irrespective of their ability level, have at least a 6.6% chance of correctly answering the question. In its present form, IRT does not presently account for student ability in assigning the probability of "guessing" the correct answer. Intuitively, we would expect a decreasing probability of guessing the correct answer with increasing item ability.

$$P(\theta) = {(1+c)}/{2} \qquad (2)$$

Figure 2 is an example of a poorly discriminating ICC. The item's ability, *b*, based on this ICC is 0.603 indicating that it discriminates between students with lower abilities than the question shown in Figure 1. However, the other parameters indicate this question was not well constructed. The small slope, *a*, of 0.399 indicates that it poorly discriminates. Furthermore, this ICC has a guessing parameter, *c*, of 0.500 indicating that all students have a fifty percent chance of guessing the item correctly. Theoretically, guessing parameters can range from 0.0 to 1.0. However, *c* values greater than 0.35 are undesirable due to the high likelihood of students successfully answering the question regardless of their knowledge (*12*).

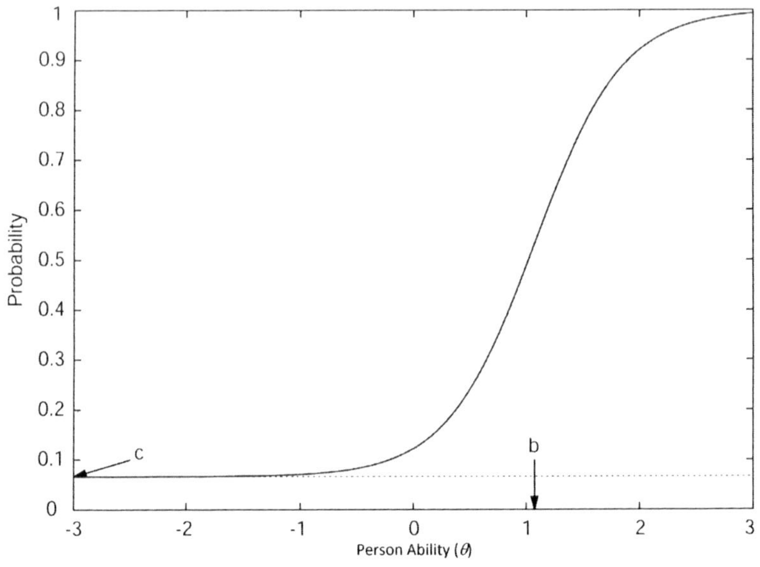

Figure 1. Ideal item characteristic curve. This item characteristic curve has a slope of 2.566 which is the item's a value. This item has an ability of 1.074 (b) with a guessing factor (c) of 0.066, indicated by the asymptote. The c value indicates that 6.6% of the students guessed the correct answer. Reproduced with permission from Schurmeier et al. J. Chem. Educ. 2010, 87(11), pp. 1268-1272. Copyright American Chemical Society.

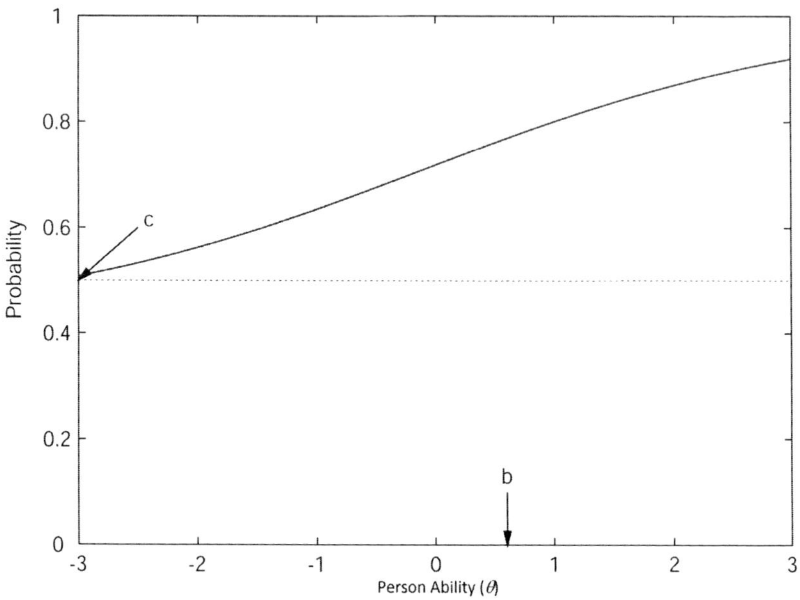

Figure 2. Poorly discriminating item characteristic curve. This ICC has a slope (a) of 0.399, an ability (b) of 0.603, and (c) value of 0.500.

Test Reliability, Unidimensionality, and Validity

For an assessment to successfully determine student subject knowledge it must meet at least two criteria: reliability and validity. Reliability refers to the test being self-consistent and capable of reproducing the same result. In other words, a given group of students taking the exam multiple times will always earn the same grade if the exam is self-consistent. Two common CTT measures of test reliability are the KR-21 measure and the Standard Error of Measure (SEM). KR-21 gives a relative measure of reliability on a scale of 0.0 to 1.0 with reliability increasing with the KR-21 value. Reliable college examinations usually have KR-21 values of 0.6 to 0.8 or higher (*58*). SEM indicates how reliably the test will reproduce the same score for a given student. In other words, if a student took a given exam multiple times, they should earn the same grade plus or minus the SEM each time. Table 1 lists KR-21 and SEM values for all of the tests discussed in this chapter. One problem with KR-21 values is that they measure test reliability relative to the test mean (*58*).

Table 1. SEM and KR-21 reliability for each individual examination given from the fall of 2004 to the spring of 2008

	KR-21	*SEM*
Fall 2004		
Exam 1	0.976	1.88
Exam 2	0.981	2.44
Exam 3	0.966	2.60
Final Exam	0.847	3.72
Spring 2005		
Exam 1	0.966	2.02
Exam 2	0.978	1.87
Exam 3	0.978	1.78
Final Exam	0.841	3.84
Fall 2005		
Exam 1	0.975	2.00
Exam 2	0.978	2.28
Exam 3	0.968	2.14
Final Exam	0.810	3.85
Spring 2006		
Exam 1	0.970	2.10
Exam 2	0.983	1.62
Exam 3	0.979	1.59
Final Exam	0.811	3.79
Fall 2006		
Exam 1	0.973	2.23
Exam 2	0.981	2.13
Exam 3	0.969	2.09
Final Exam	0.823	4.00
Spring 2007		
Exam 1	0.974	2.23
Exam 2	0.971	1.85
Exam 3	0.982	1.68
Final Exam	0.773	3.55

Continued on next page.

Table 1. (Continued). SEM and KR-21 reliability for each individual examination given from the fall of 2004 to the spring of 2008

	KR-21	*SEM*
Fall 2007		
Exam 1	0.979	2.05
Exam 2	0.982	2.12
Exam 3	0.973	2.19
Final Exam	0.830	3.99
Spring 2008		
Exam 1	0.982	1.93
Exam 2	0.984	2.07
Exam 3	0.984	1.84
Final Exam	0.838	3.63

One IRT reliability measure is the Reliability Index (or item-trait correlation) which is a biserial correlation between ability and the item. In effect, this is a measure of the discrimination ability of all the items in a given examination or set of examinations. Reliability indices can have values between 0 and 1 with the most reliable examinations having values close to 1. As can be seen from Table 2, across the entire academic year our tests are reliably discriminating students over the entire ability range.

Table 2. Reliability index of academic year examinations from the fall 2004 to the spring 2008 calculated using BILOG-MG 3

Academic Year	*Reliability Index*
Fall 2004 – Spring 2005	0.979
Fall 2005 – Spring 2006	0.990
Fall 2006 – Spring 2007	0.990
Fall 2007 – Spring 2008	0.989

Unidimensionality must be present for exams to be assessed using IRT. This indicates that the students' responses to the test items are a function of the students' abilities and that an underlying trait (in this context general chemistry) is associated with their responses. The unidimensionality of the response data permits placement of students on the ability scale (*9*). To show unidimensionality is present in the examinations, a correlation matrix was generated using the statistics program SAS 9.1.3. Visual examination of the item pair correlations generated by SAS 9.1.3 showed a positive manifold between questions (*16*). This method demonstrates that the same trait was being measured in the examinations, which in this case is general chemistry.

Validity indicates whether the exam assesses the topics that it intends to test. There is no single measure of test validity, but several factors are commonly examined. For example, content validity focuses on the consensus of a group of experts that an examination covers an array of topics necessary for mastery of chemical knowledge at the appropriate level. All tests reported in this chapter were reviewed by at least two faculty or staff members teaching the courses prior to their being administered.

Item and Total Information Curves

Each ICC generated using IRT has a corresponding item information curve that indicates which students, based on their ability, are best assessed by the question. These individual item information curves are then integrated into a total information curve that provides information about the reliability of the examination. Figure 3 is an example of an ICC for an item in our test database (Item 1) having an ability, *b*, of 0.723, discrimination factor, *a*, of 5.835 and guessing factor, *c*, of 0.130.

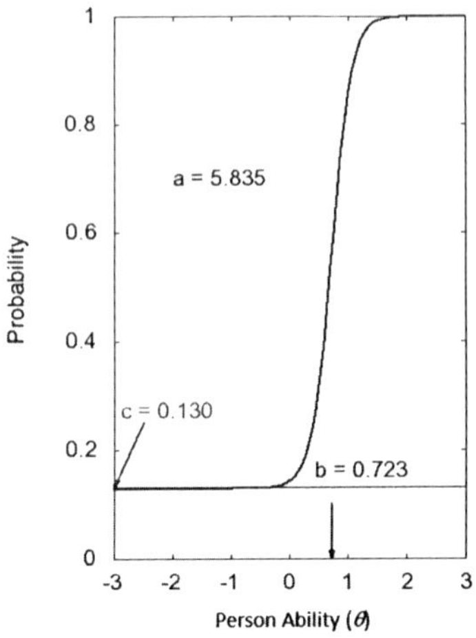

Figure 3. Item characteristic curve for a test item. Figure generated from data gathered on exam 1 fall 2003, 2004, spring 2004 and 2005. Item characteristic curve indicates that this item has an ability, b, of 0.723, asymptote, c, of 0.130 and discrimination parameter, a, of 5.835.

Item information curves are plots of ability versus gathered information for a specific exam question, and they are centered on the ability, *b*, calculated for that item and are symmetrically distributed around the item's ability. Just as a steep slope for an item characteristic curve is indicative of a highly discriminating item, a tall, thin peak on an item information curve does the same. A shorter, wider item information curve indicates greater uncertainty in the distinguished ability levels corresponding to a shallow slope in the item characteristic curve having an ill-defined midpoint. The item information curve (ICC) shown in Figure 4, correlating to the same test question from Figure 3, indicates that this item provides a great deal of information about students with an ability near 0.723; however, it gives no information about a student with an ability greater than two since they all answer it correctly. This curve will also give us no information about students with abilities less than zero because they either answered the question correctly by guessing or answered the question incorrectly.

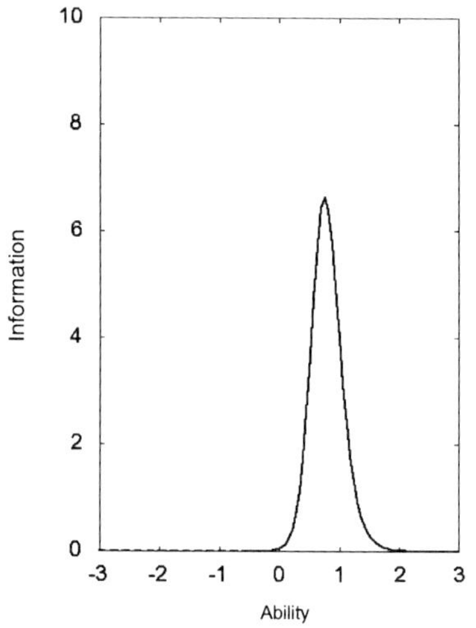

Figure 4. Item information curve generated for the question discussed in Figure 3. Figure generated from exam 1 fall 2003, 2004, spring 2004 and 2005 data.

In Bilog MG3 total information curves (TIC) are generated by integrating all of the item information curves for a single examination (or an entire year of examinations) into a single graph. Total information curves take into account the amount of information each item contributes to minimize the uncertainty of the student abilities (*9*). Figure 5 is an example of a TIC for an entire year of examinations. Exams written with a suitable range of item abilities and highly discriminating items generate a total information curve with a large centered information peak and standard errors that are small in the center and large on both ends. Embretson and Reise (*11*) provide details as to how TIC's are generated. Licensing agencies design their assessments so that a critical number of questions at the "passing" score (or person ability) level are included. This assures that examinees who fail the exam have been fairly excluded. Computerized adaptive testing, such as used in the Graduate Record Exam (GRE), is built based upon a large bank of precalibrated examination items using a TIC to insure the ability level of each examinee is estimated to a specifiable level of precision.

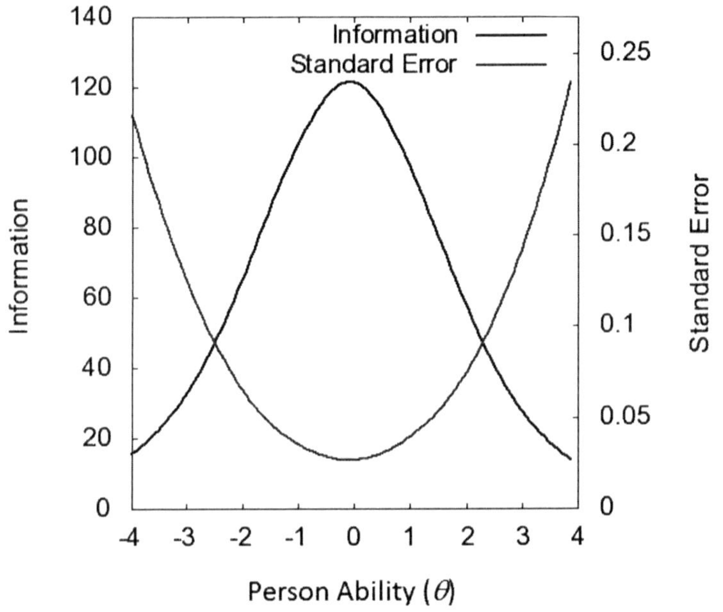

Figure 5. Total information curve generated from the fall 2006–spring 2007 academic year.

As indicated in Table 1, CTT produces a single value SEM that is supposedly applicable across the observed score range. However as shown in Figure 5, the error in a student's ability (amount of chemistry knowledge) is very small, toward the center of the graph but greater on both ends of the ability scale. This is because it is difficult to write assessment items that discriminate well between low F and F students or between A and high A students. The TIC shown in Figure 5 indicates that the students' abilities are assessed with a small amount of error across the majority of student ability levels. Even for the A and high A students (person abilities near 2), application of the standard error to their assessed person ability does not result in a change of grade level. This indicates that our assessments, improved through the use of IRT, accurately distinguish student abilities across the grade scale.

Computerized Testing

Beginning in academic year 2000-2001, computerized testing was instituted in the Freshman Chemistry program at the University of Georgia (UGA) using an in-house program, JExam, which has some features that are relevant to this chapter (*17–19*). A longitudinal database of student performance on our computerized assessments for nearly 11 years has been maintained. Much of the information in this chapter is based on IRT analysis of a database of nearly 13,000 questions containing responses from approximately 7,000 students. All midterm exams were administered using JExam while multiple choice final examinations were administered on paper so that the exam could be given to all students simultaneously.

JExam questions are not solely multiple choice. In the tables provided below are references to number entry (students enter a number into an appropriate space), text entry (similar to number entry but with text), multiple answer (similar to multiple choice but more than one answer is correct), and drop down lists (students choose from several options which appear when the list expands). Multiple response questions on a single item are commonly used. For example, three separate number entry responses may be required in a question such as, "How many protons, neutrons, and electrons are present in one atom of $^{35}P^{3-}$?" In this case the students should enter the number 15 in a box labeled "Number of protons in one atom of $^{35}P^{3-}$ =", the number 20 in the box labeled "Number of neutrons in one atom of $^{35}P^{3-}$ =", and the number 18 in the box labeled "Number of electrons in one atom of $^{35}P^{3-}$ =." If the students miss any one part of this question, IRT analysis would indicate that they had missed the problem entirely. While this is true of the IRT analysis, it is not true of the students' test grade analysis where they receive partial credit for the correct responses.

Easy General Chemistry Topics

Unsurprisingly, some topics in general chemistry are relatively easy for students. Their item abilities discriminate only between the very lowest student abilities. Virtually all students correctly answer questions on these topics: unit conversions, significant figures, balancing equations, oxidation states, reaction stoichiometry (without excess reagent), electron configurations of elements, determining empirical formulas, numbers of protons, neutrons and electrons in an atom or isotope, atomic radii trends, solution dilution problems, ideal gas law problems, interpreting phase diagrams, osmotic pressure problems, freezing point depression problems, determining ΔH_{rxn} from ΔH_f, Hess's Law and rate law calculations. Because these topics are easier for students, less in-class instructional time could be devoted to them in favor of more difficult topics.

Consistently Difficult General Chemistry Topics

A primary research goal was the use of IRT to determine those areas of a typical general chemistry curriculum that confuse a majority of students and which may be remediated by some instructional changes. In pre-existing research, those topics were determined by looking at small sample sizes, usually around 100 or fewer students (*20–33*). Our research uses approximately 1200 students each academic year, thus providing more information on the difficult topics coupled with greater statistical certainty. Furthermore, we used IRT to determine which portions of our student population, based upon their determined personal IRT ability levels, were troubled by these difficult topics. Difficult topics previously identified in the literature include: bonding, compounds in solution, equilibrium, energy, chemistry on the molecular level, and nomenclature (*21*, *25*, *34–37*).

Several very basic general chemistry ideas are taught early in the academic year. We found that if these topics are well understood by the students, they can build upon that throughout the year. One of the most basic but incredibly important ideas centers on the students' understanding of the particulate nature of matter. As Gabel and Johnstone have previously noted, chemists easily transition between the macroscopic and submicroscopic understanding of nature using chemical symbolism to develop a visual representation of the particulate nature of matter (*38*, *39*). Our research indicates that for a significant fraction of our students relating chemical symbolism to a particulate understanding of molecules posed a significant problem in their comprehension of chemistry which affected their performance in the course. Our IRT research revealed to us the magnitude of this problem for our student population.

Particulate Nature of Matter

After construction of the JExam testing system in 2000, data was collected for four years without extensive analysis, and IRT analysis of the database began in academic year 2004-2005. Question 1, Table 3 which asks students to a) type in the name as well as b) the number of ions present in one formula unit of $Al(NO_3)_3$ and several other similar questions are among the most discriminating questions in our entire database. Students that earn a low C, D or F on the exam consistently miss this question.

Previous research has indicated that many students have difficulty grasping the idea that ions have different structures than atoms and covalent compounds, particularly polyatomic ions (*24*, *25*, *40–43*). Over the next two academic years, several variations of questions on the particulate nature of matter were asked on examinations and analyzed in an attempt to elucidate where the students struggled in their understanding of this key topic. These relevant questions along with their IRT data are presented chronologically in Table 3. The IRT analysis of these questions indicated that students must earn an A, B or high C on the examination to routinely correctly identify the following: a) the number of ions in an ionic compound containing polyatomic ions, b) ionic versus covalent compounds, c) a physical property that would be predicted by the knowledge of ionic/covalent structure. Lower performing students fail to answer these questions correctly. While there are many possible reasons why this concept is so problematic for the lower ability students, the evidence is compelling that facilitating students' understanding of the particulate nature of matter is a key component to improving overall student performance in general chemistry.

Table 3. Questions and Statistics Related to Particulate Nature of Matter

Question	*Options*	*Question Type*	*b value*	*a value*	*c value*
1. How many ions are present in one formula unit of the compound shown above?	Na_2CO_3 Na_3PO_4 $Al(NO_3)_3$	Number entry	0.857 1.230 0.843	0.108 0.290 0.149	0.001 0.001 0.001
2. How many potassium, oxygen, carbon, and carbonate ions are present in one formula unit of this compound? K_2CO_3		Number entry	1.074	1.104	0.001
3. How many oxygen atoms and hydrogen atoms are present in one formula unit of Al(OH)3. How many aluminum ions, oxide ions, hydroxide ions and hydrogen ions are present in one formula unit of $Al(OH)_3$?		Number entry	1.034	0.386	0.152
4. Label each of the following as an atom, an ion, a molecule, a formula unit or none of these.	MgS O_2 Mg SO_3^{2-} Al^{3+}	Drop down lists	1.330	1.583	0.111
5. Choose all of the (*ionic or covalent*) compounds from the list below.	$Ca(OH)_2$ LiN_3 Sr_3N_2 CO_2 NI_3 CBr_4	Multiple answer	0.653	0.817	0.161

Question	*Options*	*Question Type*	*b value*	*a value*	*c value*
6. How many mL of x M $Sr(OH)_2$ are required to make y mL of a z M $Sr(OH)_2$ solution? What is the molar concentration of the Sr^{2+} ions in the z M $Sr(OH)_2$ solution? What is the molar concentration of the OH^- ions in the z M $Sr(OH)_2$ solution?	x, y, and z are variables appropriate for this problem. Ca(OH)$_2$ and Ba(OH)$_2$ were used in question variations.	Number entry	1.325	1.212	0.001
7. There are 20.0 drops of solution in 1.00 mL of solution. How many bromide, Br^{1-}, ions are present in x drops of y M $MgBr_2$?	x and y are variables appropriate for this problem.	Number entry	1.660	1.180	0.001
8. What mass of $FeCl_3$ would contain the same **total** number of ions as 16.8 g of $Al_2(SO_4)_3$?		Multiple choice	1.949	1.145	0.120
9. Which **aqueous** solution would have the **lowest** vapor pressure at 25°C?	1 m NaCl 1 m Na_3PO_4 1 m $MgCl_2$ 1 m $C_6H_{12}O_6$ 1 m $C_{12}H_{22}O_{11}$	Multiple choice	1.857	1.207	0.149

Molecular Polarity and Intermolecular Forces

Previous researchers determined that students struggle with the concept of molecular polarity because they have specific misconceptions including the notion that individual atoms have polarity (*25*, *44*). Some students cannot combine the concept of electronegativity with that of polarity (*25*, *45*) while other students have trouble visualizing molecular shapes (*23*, *45*). Understanding molecular polarity is a necessary, but not sufficient, condition to apprise the intermolecular forces present in the liquid and solid states of a molecule. Consequently, it is not surprising that students struggle with both polarity and intermolecular forces (*24–26*, *28*, *46*). It has also been shown in earlier research that some students believe that intermolecular and intramolecular attractions are the same (*i.e.* they believe that covalent bonds and intermolecular attractions are one and the same) (*26*, *28*, *47*). Over time, we have tried to isolate which of these misconceptions have the greatest impact on our student population. Presented in Table 4 are several questions developed in attempts to assess specific knowledge about students' understanding of molecular polarity and intermolecular forces.

For the students in this study, a direct question regarding which molecules are polar or nonpolar (Table 4, Question 1) discriminated only between students who earned a high or low F on exams. But combining the ideas of bond and molecular polarity (Table 4, Question 2) makes the question more difficult; indeed, only A and very high B students correctly answered it. However, as demonstrated later in the discussion regarding acid strengths and solution concentrations, question phrasing that places two seemingly contradictory ideas together is problematic for students. In this case, the contradictory phrasing is a nonpolar molecule with polar bonds.

Question 4 in Table 4 was answered correctly only by students earning high A's. Notice that this question again requires students to combine concepts to achieve a correct answer. In this case, they must address the fact that all molecules have London dispersion forces, but some molecules also exhibit other intermolecular forces.

In an attempt to remove the possible difficulty of drawing Lewis structures, the students were asked a question about intermolecular forces with both the molecular images and the intermolecular forces provided (Question 6 in Table 4). Three of the five image questions were extremely easy for students. However, two of the items were significantly more difficult than the other three. The molecular images used in the question are shown in Figure 6. The two difficult questions used images 1 and 2 which ask the students to identify London dispersion forces. Many A ability students were unable to answer these correctly. Since both of these questions contained images of nonpolar molecules with an induced dipole, we hypothesize that many students focused on the indicated attraction and assumed it was a dipole-dipole intermolecular attraction without looking at the molecular images of nonpolar molecules. Additionally, this problem was likely difficult due to the large number of mental steps (m-demand or mental capacity) needed to correctly answer the question. Students needed to identify and separate the individual molecules, distinguish between the intramolecular bonds and the intermolecular attractions, determine whether the molecules were polar or non-polar, reconcile the polarity and partial charges as temporary, induced or permanent dipoles, and then identify the primary intermolecular force based on the sum of those pieces of information.

To determine if the students struggled with drawing the molecules given, determining if the molecule is polar, or both, students were asked Question 10 in Table 4. This question requires students to recognize that polar molecules are water soluble and then choose the polar molecule from a series of pentavalent phosphorus compounds. Only the highest ability students correctly answered this question. However, the guessing factor is quite high indicating that many of our students eliminated at least one of our distracters.

We conclude that molecular polarity and intermolecular forces are very difficult topics for students to understand. Any time students are required to determine molecular polarity or intermolecular forces based upon molecular polarities, this topic will almost certainly discriminate the high ability students from the rest. The only method found to simplify this topic to the point that lower ability students correctly determine the forces is if molecular images with intermolecular forces other than London dispersion are used.

Understanding Quantum Numbers

Electronic configurations of the elements, the basic rules of how the four quantum numbers are related to one another, and the physical interpretations that can be applied to each of the quantum numbers are taught simultaneously in our course as in most general chemistry courses. While atomic electron configurations were quite easy for students in this study, quantum number related topics were very difficult. Other researchers have also found that many students can remember the rules for quantum numbers, i.e. n, ℓ, m_ℓ and m_s; however, the physical meaning associated with these numbers tends to elude them (*22*, *25*, *48–50*). This is symptomatic of the recurring theme throughout chemistry that students place stronger emphasis on memorization than garnering physical understanding.

Given below in Table 5 are 9 questions that were administered to students during the academic years from 2005 to 2008. The first question (Question 1 in Table 5) asked the students to use their knowledge of quantum numbers, as well as their understanding of the aufbau principle and Pauli exclusion. Only the A, B and C students answered this item correctly. As soon as any application of the quantum number rules (Questions 2 – 7 and 9) were asked, only the most able students could correctly answer those questions. Even asking the students to apply quantum numbers to an image of an atomic orbital was difficult for most students (Questions 4 and 7). Question 8 was asked on a final examination after the students had done poorly on an earlier mid-term examination containing quantum number problems. The significantly lower question ability for problem 8 indicates that in the intervening time between the mid-term and final exams, the lower ability students either learned how to interpret quantum numbers and their associated physical characteristics as well as electron configurations, or they memorized these associations.

Table 4. Questions and Statistics Related to Molecular Polarity and Intermolecular Forces

Question	*Answer Options*	*Question Type*	*b value*	*a value*	*c value*
1. Which of the following molecules are nonpolar?	CCl_4 CH_2Cl_2 CH_3Cl $CHCl_4$ SiH_2Cl_2	Multiple answer	1.898	−0.485	0.192
2. Choose from this list of molecules the one that is nonpolar but contains polar covalent bonds.	NH_3 H_2Te $SOCl_2$ (S is the central atom) $BeBr_2$ HF	Multiple choice	1.456	1.522	0.217
3.Choose the dominant (strongest) intermolecular attraction in the liquid state for each of the substances listed below.	strontium sulfide, Sr_2S dimethyl ether, $H_3C–O–CH_3$ methyl amine, $CH_3–NH_2$ carbon tetrachloride, CCl4	Drop down lists	0.889	1.377	0.132
4. Which response correctly identifies all the interactions that might affect the properties of BrI?	•Dispersion force, ion-ion interaction •Hydrogen bonding force, dispersion force •Permanent dipole force •Permanent dipole force, dispersion force •Dispersion force	Multiple choice	1.038	2.828	0.293

Question	*Answer Options*	*Question Type*	*b value*	*a value*	*c value*
5. Which bond would have the smallest intrinsic bond dipole moment?	H–Br H–F H–I H–O in H_2O H–Cl	Multiple choice	1.057	1.142	0.160
6. The intermolecular force depicted in this representation is: ion–ion attraction, dipole–dipole attraction, hydrogen bonding, or London dispersion forces.	See Figure 6	Multiple choice	1.149 1.183 0.678 0.678 0.706	3.035 1.934 −1.712 −1.774 −2.667	0.125 0.114 0.004 0.004 0.004
7. Which of the following intermolecular forces is associated with ALL types of compounds?	Ion–ion interactions London Forces Covalent bonding Dipole–dipole interactions Hydrogen bonding	Multiple choice	0.640	0.057	0.211
8. Which of the following interactions is the **strongest** type of intermolecular force?	Ion–ion interactions London Forces Hydrogen bonding Dipole–dipole interactions Dispersion forces	Multiple choice	0.839	−1.605	0.182

Continued on next page.

Table 4. (Continued). Questions and Statistics Related to Molecular Polarity and Intermolecular Forces

Question	*Answer Options*	*Question Type*	*b value*	*a value*	*c value*
9. Consider the following pairs of liquids. Which response contains all the pairs that are miscible and none that are immiscible?	•Benzene, C_6H_6, and hexane, C_6H_{14} •Water and methanol, CH_3OH •Water and benzene, C_6H_6	Multiple choice	1.734	1.176	0.319
10. A sample of PBr_3I_2 dissolves in water. The best three-dimensional representations of five potential structures are shown below. Which is the most likely structure for this sample?	See Figure 7	Multiple choice	1.041	2.413	0.284

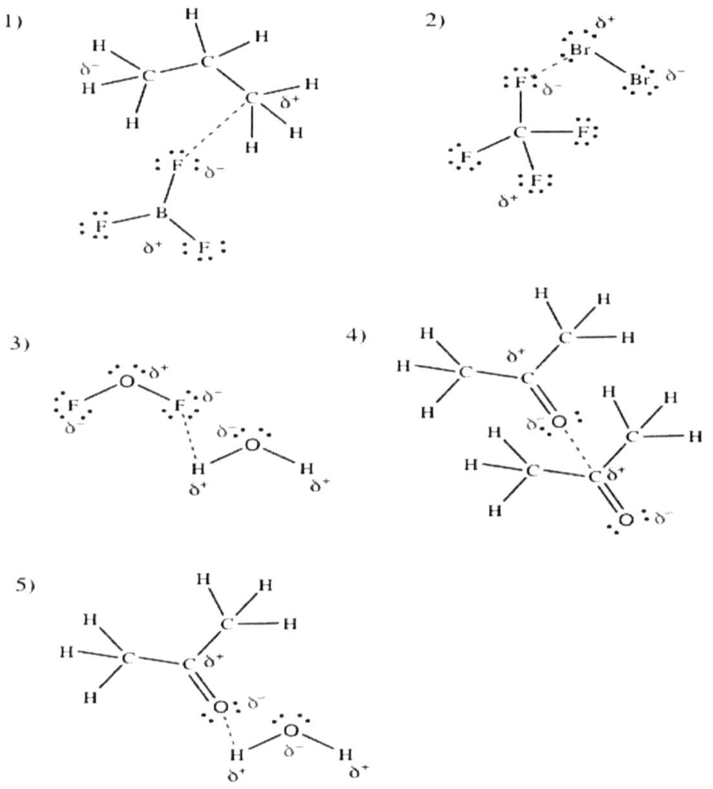

Figure 6. Molecular Images Used in Question 6 of Table 4.

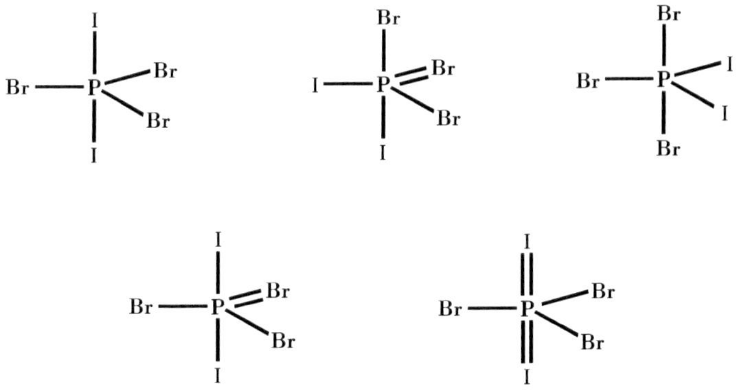

Figure 7. Molecular Images Used in Question 10 of Table 4.

Table 5. Questions and Statistics Related to Understanding Quantum Numbers

Question	*Answer Options*	*Question Type*	*b value*	*a value*	*c value*
1. Choose the set of quantum numbers which would not be correct for any of the electrons in the ground state configuration of the element S.	$n = 3$, $\ell = 2$, $m_\ell = 0$, $m_s = +1/2$ $n = 2$, $\ell = 1$, $m_\ell = -1$, $m_s = -1/2$ $n = 3$, $\ell = 0$, $m_\ell = 0$, $m_s = +1/2$ $n = 2$, $\ell = 0$, $m_\ell = 0$, $m_s = -1/2$ $n = 1$, $\ell = 0$, $m_\ell = 0$, $m_s = -1/2$	Multiple choice	1.520	0.387	0.250
2. Which quantum number describes the specific orbital within a subshell that an electron occupies?	n ℓ m_ℓ m_s	Multiple choice	1.025	1.650	0.148
3. How many d electrons are there in the ground state electron configuration for the element Fe? What is the value of the *n* quantum number for the d electrons in Fe? How many of the d electrons in Fe have +1/2 spins? How many of the d electrons in Fe have −1/2 spins?		Number entry	1.519	1.869	0.073
4. What is the maximum number of electrons in an atom that can be described by the following quantum numbers? $n = 4$, $\ell = 2$. Which image would best represent one of the orbitals the electrons described by $n = 4$ and $\ell = 2$ would be in?	See Figure 8	Number entry and Multiple choice	0.828	2.157	0.001

Question	*Answer Options*	*Question Type*	*b value*	*a value*	*c value*
5. What is the maximum number of electrons in an atom that can be described by the following quantum numbers? $n = 3$, $\ell = 1$.		Multiple choice	0.975	2.462	0.184
6. In the iron atom, how many electrons are in orbitals for which the angular momentum quantum number is 0?	2 8 6 10 26	Multiple choice	1.277	2.086	0.207
7. Which quantum number distinguishes between the three p orbitals shown below?	The principle quantum number. The magnetic quantum number. The angular momentum quantum number. The spin quantum number. None of the quantum numbers. See Figure 9	Multiple choice	0.956	1.553	0.133
8. What is the maximum number of electrons that can be described by the following set of quantum numbers? $n = 5$, $\ell = 3$, $m_\ell = -2$.	48 40 10 2 1	Multiple choice	1.607	0.413	0.122
9. How many electrons can be described by the $n = 4$ and $\ell = 2$ quantum numbers?		Number entry	0.828	2.157	0.001

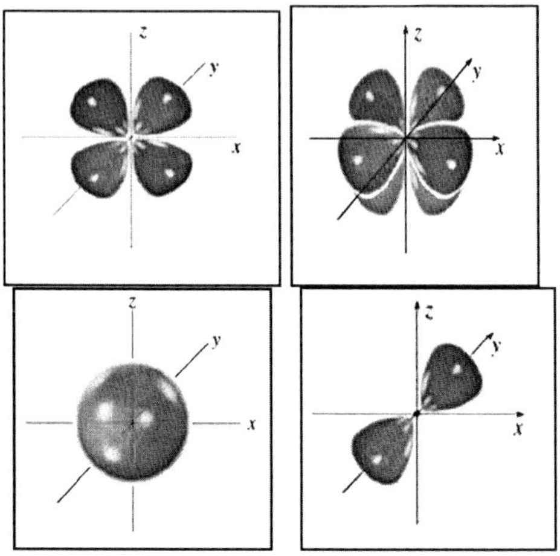

Figure 8. Answer Options Used in Question 4 of Table 5

Students could also choose the response "Not enough information given to answer the question."

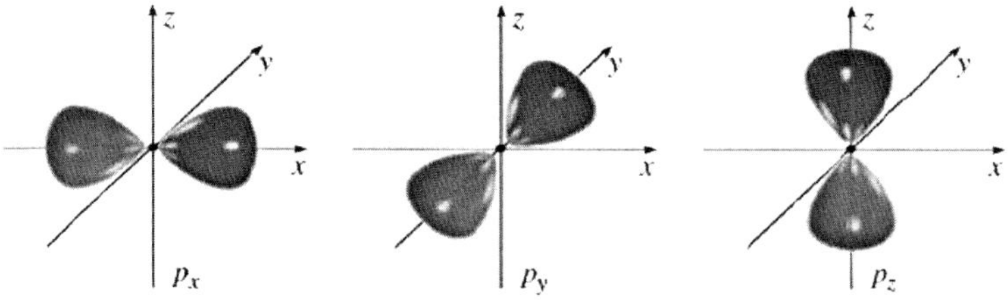

Figure 9. Orbital Images Used in Question 7 of Table 5

Use of the Terms "Strong, Weak, Concentrated and Dilute" in Relation to Acidic or Basic Solutions

Unfortunately for students, chemists frequently use common English words or terminology but with a different and highly specialized connotation. This is certainly true when it comes to the terms strong, concentrated, weak, and dilute. For chemists, strong and weak refers to the behavior of electrolytes. Weak and dilute refer to relative concentrations of solutions. This can present a serious problem for students (*40*). It is not uncommon for students to believe that strong and concentrated are equivalent terms or that weak and dilute connote the same idea. Gabel pointed out that it is common to say such things as "the coffee is strong" instead of the more correct "the coffee is concentrated" (*38*).

To assess this topic, questions were written that specifically addressed these commonly confused terms. Almost all students could correctly pick a weak or strong acid from a list of molecules (Questions 1 and 2 from Table 6). Essentially 85% of the students correctly responded to these two problems. However, if asked specifically about the relationship of the four terms (Question 3), the question was increasingly difficult. Only students with the highest ability answered it correctly. Repeated conversations with students indicated that B and lower ability students did not understand that strong, weak, and nonelectrolytes can all form concentrated solutions. A significant fraction believed that only strong electrolytes can form concentrated solutions.

When the examination question changed from understanding to applying the phrases to molecular images (Question 4), it was found that students struggled more with their molecular understanding of the term "dilute." The most difficult combinations occurred when the students were asked to identify the dilute/weak acid and dilute/strong acid combinations. This is clearly one area of the general chemistry curriculum requiring special emphasis in classroom instruction.

Table 6. Questions and Statistics Related to Use of the Terms "Strong, Weak, Concentrated and Dilute" in Relation to Acidic or Basic Solutions

Question	*Answer Options*	*Question Type*	*b value*	*a value*	*c value*
1. Which one of the following is a strong acid?	HI HF HNO_2 HCN HBrO	Multiple choice	1.580	−1.580	0.189
2. Which one of the following is a weak acid?	$HClO_4$ HCl HBr HI CH_3COOH	Multiple choice	1.124	−1.849	0.190
3.Select the substance(s) from the choose box at the right that has the property listed on the left. There is only one correct answer per property. •Conducts electricity well in dilute aqueous solutions. •Ionizes only slightly in dilute aqueous solutions. •Can form **concentrated** solutions. •Does **not** conduct electricity in aqueous solutions. •Can form **dilute** solutions.	HCl HF CH_3OH HCl and HF HCl, HF, and CH_3OH	Drop down lists	1.681	2.486	0.041
4. Choose the picture that is the best representation of a **concentrated strong acid in solution**. For the sake of clarity, the dissolving water molecules have been removed from these pictures. The legend for the pictures is given immediately below. (Other questions of this type asked the students to identify concentrated weak, dilute strong, and dilute weak acid solutions.)	See Figure 10	Drop down lists	0.977 0.967 0.703 0.953	−0.053 0.017 1.086 0.977	0.213 0.256 0.237 0.186

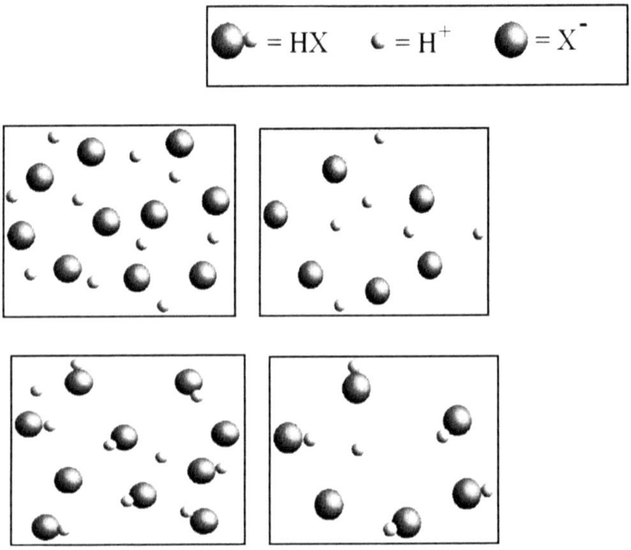

Figure 10. Molecular Images Used in Question 4 of Table 6

Molecular Image Problems

As indicated above, one of the most significant teaching challenges is facilitating students' understanding and interpretation of chemical symbolism. Consequently, it was not surprising that problems based upon molecular images were challenging for most students. Many of the topics taught in the general chemistry curriculum can be assessed with incorporated molecular images. Other studies have shown similar results (*31*, *35*, *40–42*, *51*, *52*).

Students were asked basic questions regarding physical changes, stoichiometry, equilibrium, or kinetics. Only the lowest ability students were unable to answer them correctly; this is shown by Question 1 in Table 7. When molecular images were added to the problem, the difficulty increased and the question became more challenging for students to answer correctly (Questions 2, 4, 5, 6, and 8). Lower ability students struggled with correctly answering an effusion problem that asks the students to rank molecules or atoms that would effuse the fastest, based upon molecular images (Question 3). Question 7 was easier for students, likely because the question gave hints and prompts to make them think conceptually about why they were choosing a specific image. For example, two of the four image sets depicted chemical changes that might not have distracted the low ability students. Question 6 also required students to remember the solubility rules in order to choose the correct answer. Students had to recognize that $BaBr_2$ ionizes in water yielding aqueous Ba^{2+} ions and $2Br^{1-}$ ions not $BaBr_2$ or Br_2, both of which were distracters.

Table 7. Questions and Statistics Related to Molecular Image Problems

Question	*Answer Options*	*Question Type*	*b value*	*a value*	*c value*
1. Which answers are chemical changes and not physical changes?	•Freezing of water •Rusting of iron •Dropping a piece of iron into hydrochloric acid (hydrogen gas is produced) •Burning a piece of wood •Emission of light by a kerosene oil lamp	Multiple answer	0.636	0.501	0.191
2. The compound H_4P_2 boils at 57.5 °C. Choose the picture that best represents what molecular H_4P_2 looks like after boiling has occurred. A purple ball represents a phosphorous atom, and a white ball represents a hydrogen atom.	See Figure 11	Multiple choice	0.622	2.572	0.243
3. Rank the following atoms and molecules based on which would effuse the fastest. A red ball represents an oxygen atom, a black ball represents a carbon atom, and a white ball represents a hydrogen atom. Assume all are gases at the same temperature.	See Figure 12	Drop down lists	1.291	0.854	0.283
4. The reaction occurring in these images is R $\rightleftharpoons$ B, where R represents a red ball and B represents a blue ball. Using the following images at what time is equilibrium first established?	See Figure 13	Multiple choice	1.037	1.465	0.189

Question	*Answer Options*	*Question Type*	*b value*	*a value*	*c value*
5. Depicted below is a reaction vessel containing a mixture of H_2 and Cl_2 gas (before reaction begins). The green spheres represent chlorine atoms, and the white spheres represent hydrogen atoms. Chlorine and hydrogen react to form hydrochloric acid, HCl. $Cl_2(g) + H_2(g) \rightarrow 2HCl(g)$ 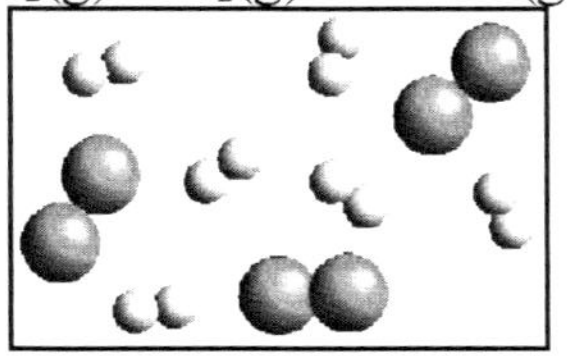 Choose the image below that best represents the reaction vessel if the reaction goes to completion.	See Figure 14	Multiple choice	0.926	1.450	0.197

Continued on next page.

Table 7. (Continued). Questions and Statistics Related to Molecular Image Problems

Question	*Answer Options*	*Question Type*	*b value*	*a value*	*c value*
6. Depicted below are 2 vessels: The one on the left containing Ba(OH)2 and the other containing HBr (before reaction begins). The blue spheres represent bromide atoms, the white spheres represent hydrogen atoms, the grey spheres represent barium atoms, and the red spheres represent oxygen atoms. When a reaction between these two reagents takes place, what products are formed? 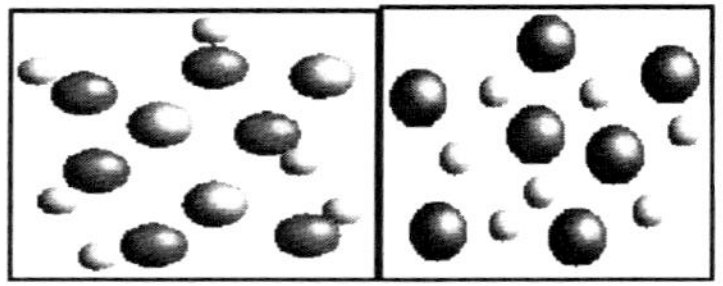	See Figure 15	Multiple choice	1.249	1.530	0.001

Question	*Answer Options*	*Question Type*	*b value*	*a value*	*c value*
7. The series of images below depict chemical or physical changes. The starting image is on the left and the ending image is on the right. Select the image set that could depict the propane gas, C_3H_8, condensing to a liquid.	See Figure 16	Multiple choice	0.969	−1.539	0.209
8. Consider the first-order reaction A → B in which A molecules (blue) are converted to B (red) molecules. Given the above pictures at t = 0 min and t = 4 min, how many A and how many B molecules will be present at t = 2 minutes? How many A and how many B molecules will be present at t = 8 minutes? What is the half-life of the reaction? 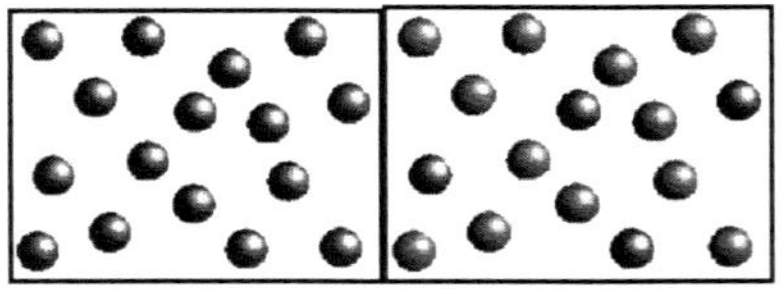 t = 0 min t = 4 min		Number entry	1.058	1.023	0.125

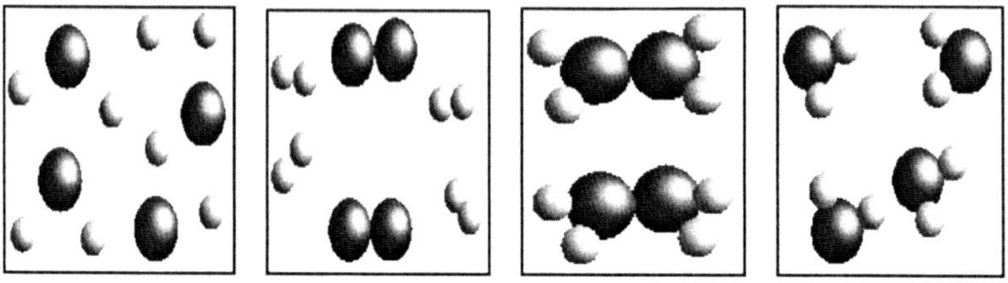

Figure 11. Molecular Images Used in Question 2 of Table 7

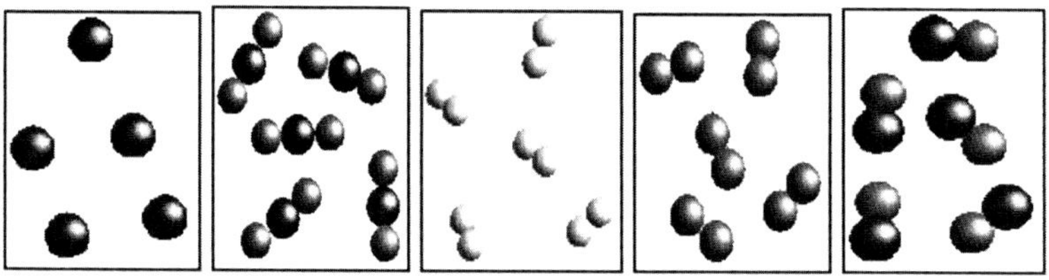

Figure 12. Molecular Images Used in Question 3 of Table 7

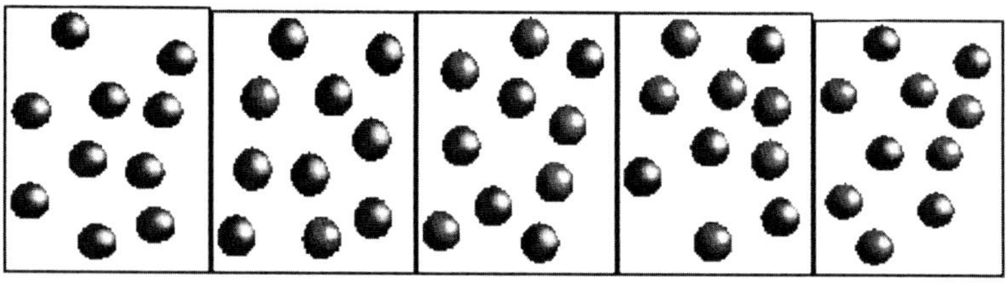

t = 0.0 s t = 5.0 s t = 10.0 s t = 15.0 s t = 20.0 s

Figure 13. Molecular Images Used in Question 4 of Table 7

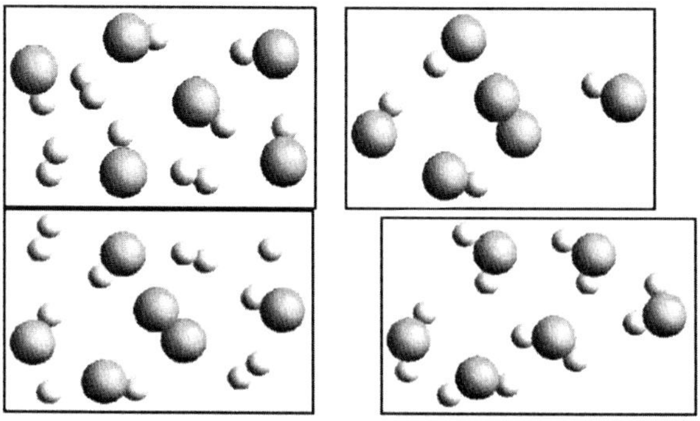

Figure 14. Molecular Images Used in Question 5 of Table 7

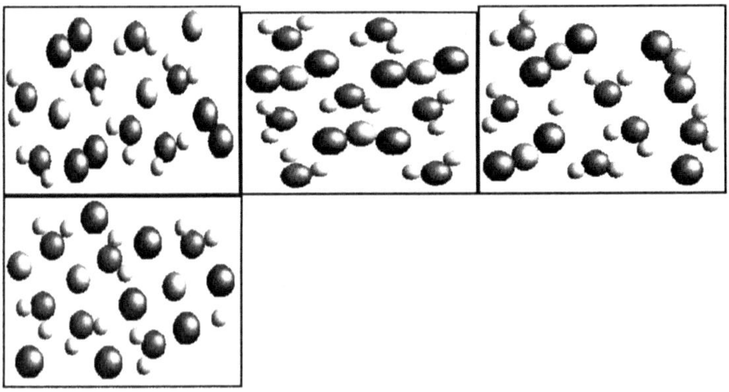

Figure 15. Molecular Images Used in Question 6 of Table 7

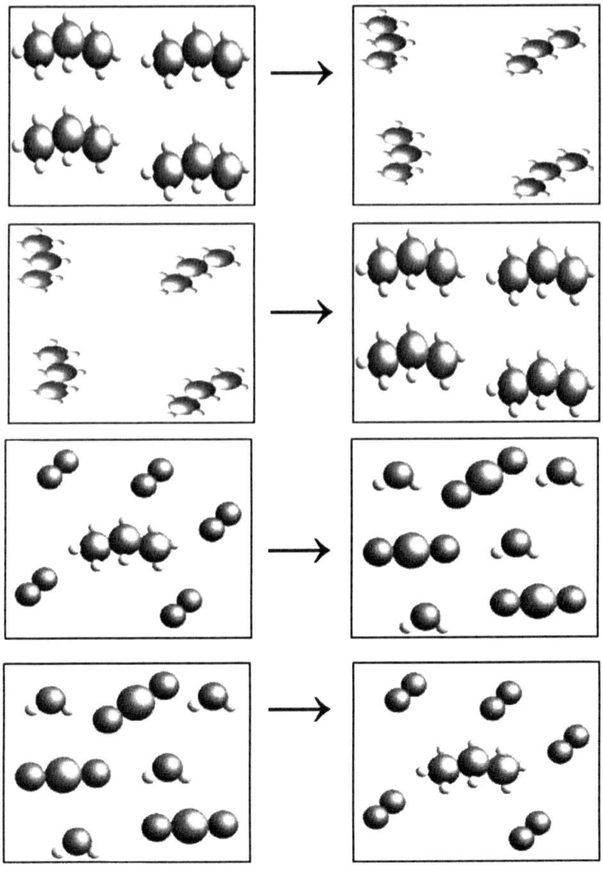

Figure 16. Molecular Images Used in Question 7 of Table 7

The Mole Concept

Many students have a relatively small cognitive understanding of the mole concept. While most students can easily work algorithmic problems that address the mole concept, most students struggle with the underlying concepts of these problems. (*53–56*). Some conceptual problems were developed that indicate where their conceptual problems lie. Question 1 of Table 8 reemphasizes the previously noted difficulty students have with their understanding of atoms, ions, and molecules. Conversations with students indicated that they have little trouble answering the first two parts of this question. Rather, they struggled with deciding whether the number of atoms and molecules or the numbers of atoms are equal in the two species. This idea was very difficult for students to grasp; indeed, only the most able A and B students answered this question correctly.

Question 2 is a multiple step problem that may well exceed the working memory of lower ability students. We estimate that there are at least three necessary cognitive steps necessary before the students use the mole concept to answer this question: 1) recognition that the masses of both samples are the same, 2) understanding that each sphere can represent a mole, so there are more moles on the right, 3) use of the definition of molar mass to recognize that to have the same mass with fewer moles means that the mass of each mole is greater. This question discriminates between high and low C students, but not well. It was hypothesized that this problem would have a higher ability; however, an item very similar to this was given to the students on a practice examination. Their familiarity with a similar problem likely affected the IRT values.

Solution Calorimetry

Even though the students performed several solution calorimetry experiments during the course of the semester and the material was covered in lecture, this topic was still difficult for lower ability students. This is in accord with the findings of others (*57*). As seen from the question (Table 9), when asked to determine the heat flow, moles of reaction, and ΔH for the reaction of aqueous $PbBr_2$ and NaCl solutions, only A and B students answered this item correctly. Essentially the same results were achieved if aqueous NaOH and HCl solutions were used instead of $PbBr_2$ and NaCl.

Table 8. Questions and Statistics Related to the Mole Concept

Question	*Answer Options*	*Question Type*	*b value*	*a value*	*c value*
1. The density of Mg is 1.74 g/mL. The density of H_2 is 8.9×10^{-5} g/mL. In the following responses compare 1.0 mole of Mg to 1.0 mole of H_2.	The mass of 1.0 mole of Mg is [***drop down list***] the mass of 1.0 mole of H_2. The volume of 1.0 mole of Mg is [***drop down list***] the volume of 1.0 mole of H_2. The number of atoms in 1.0 mole of Mg is [***drop down list***] the number of molecules in 1.0 mole of H_2. The number of atoms in 1.0 mole of Mg is [***drop down list***] the number of atoms in 1.0 mole of H_2.	Drop down lists where the students choose from these options: $<$ $>$ $=$	1.847	1.419	0.102
2. Look at the following image and determine which element on the left or right has the greater molar mass. 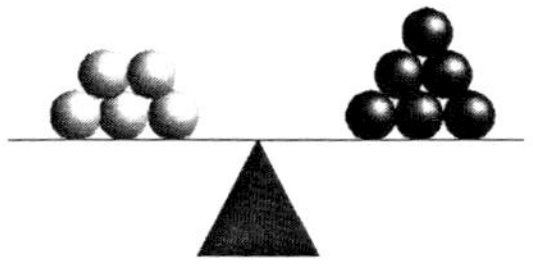	•Right •They have the same molar mass. •There is not enough information to answer. •Left •None of the above	Multiple choice	0.626	0.818	0.186

Table 9. Questions and Statistics Related to Solution Calorimetry

Question	*Answer Options*	*Question Type*	*b value*	*a value*	*c value*
1. A coffee-cup calorimeter having a heat capacity of 350.0 J/°C is used to measure the heat evolved in this aqueous reaction. $PbBr_2(aq) + 2\ NaCl(aq) \rightarrow PbCl_2(s) + 2\ NaBr(aq)$ 200.0 mL of 0.300 M $PbBr_2$ are mixed with 500.0 mL of 0.300 M NaCl, both solutions are initially at 20.000 °C. After thorough mixing, the temperature of the mixture is 20.370 °C. Assume that the solutions have a density of 1.00 g/mL and a specific heat of 4.18 J/g °C. Find the amount of heat, in kJ, liberated or absorbed in this sample. Remember to use + or − signs! How many moles of reaction were consumed in this sample? What is the ΔH for this reaction? Remember to use + or − signs!	q = [number entry] n = [number entry] ΔH = [number entry]	Number entry	0.902	1.240	0.001

Inorganic Nomenclature

The IRT analysis identified several inorganic nomenclature subcategories that highly discriminated between students' ability levels. As the nomenclature rules for a given compound became more complex, not surprisingly, the questions became ever more discriminating between higher-level students. For example, exam results reveal that some students who earned F's on the exam could correctly answer questions about ionic compounds with transition metal cations (Question 1 in Table 10). However, both D and F ability students failed to correctly name binary covalent compounds (Question 2). Essentially all of the A students could correctly name ternary acids such as HIO_2 (Question 3), whereas only our highest A ability students could name ternary acid salts (Question 4). These results suggest that increasing cognitive load may determine student performance on inorganic nomenclature problems.

Table 10. Questions and Statistics Related to Inorganic Nomenclature

Question	*Question Type*	*b value*	*a value*	*c value*
1. What is the correct name of this chemical compound? **$CoBr_3$** Use the modern naming system.	Text entry	1.295	−0.420	0.001
2. What is the correct name of this chemical compound? **BrF_5**	Text entry	0.920	0.431	0.001
3. What is the correct name of this chemical compound? **HIO_2 (aq)**	Text entry	1.195	1.666	0.001
4. What is the correct chemical name for this compound? **RbH_2AsO_4**	Text entry	1.005	2.854	0.001

Conclusion

General chemistry topics that are difficult for UGA students were identified using IRT. Previous research has shown these topics to be universally difficult for all general chemistry students, not just for students at UGA (*24*, *26*, *35*, *46–51*, *53*). However, previous research has not informed instructors which of these topics were pervasively difficult throughout the ability levels of the student population. This research correlated topic difficulty to the grades students earned on exams. For example, we found that topics such as molecular level understanding and the meaning of quantum numbers are difficult for all UGA students, except those who earn A's on exams. Other topics, such as understanding the difference between an ion, atom, or molecule and calorimetry are difficult for only C, D and F level students.

This study also elucidated the core chemistry topics that initially are difficult for students and require emphasis in instruction at the beginning of the course. For example, understanding the structure of ionic compounds (particularly polyatomic ions) is necessary to fully understand the number of ions dissolved in solutions, a concept that impacts the following topics: acids and bases; strong, weak and non-electrolytes; freezing point depression; boiling point elevation and vapor pressure lowering. Understanding of bond polarity versus molecular polarity is also difficult for many students, and is essential to understanding intermolecular forces. We have found that helping our students understand these core topics greatly improves success in the remainder of the course.

Recognition of which ability students do not grasp certain topics can help instructors plan their lessons accordingly to emphasize specific topics. Additionally, with knowledge of which topics are easy for all students, instructors can feel more comfortable covering them in less detail and time in their classes. While IRT has proven to be very informative about topic difficulty, future analysis and research, such as student interviews, are necessary to determine why students struggle with these topics.

Based upon this knowledge of difficult topics, we are working to identify students that are on the verge of grasping the material using their performance on assigned homework. Once identified, these "at risk" students have been invited to interventional help sessions to discuss the specific, difficult topic in a more innovative way that should lead to a better understanding of these topics. Initial results of this work are promising and will be presented in a forthcoming publication.

References

1. Martin, R. The Laws of the Jungle Also Apply in College. *The Athens Banner-Herald*, October 23, 2005.
2. Emons, W. H. M.; Meijer, R. R.; Denollet, J. *J. Psychosom. Res.* **2007**, *63*, 27–39.
3. Bhakta, B.; Tennant, A.; Horton, M.; Lawton, G.; Andrich, D. *BMC Med. Educ.* **2005**, *5*.
4. Scherbaum, C. A.; Finlinson, S.; Barden, K.; Tamanini, K. *Leadership Q.* **2006**, *17*, 366–386.
5. Cohen, A. S.; Bottge, B. A.; Wells, C. S. *Council Exceptional Children* **2001**, *68*, 23–44.
6. Lawson, D. M. *J. Manipulative Physiol. Ther.* **2006**, *29*, 393–397.
7. Wagner, T. A.; Harvey, R. J. *Psychol. Assess.* **2006**, *18*, 100–105.
8. Puhan, G.; Gierl, M. J. *J. Cross Cult. Psychol.* **2006**, *37*, 136–154.
9. de Ayala, R. J. *The Theory and Practice of Item Response Theory*; The Guilford Press: New York, 2009.
10. Zimowski, M.; Muraki, E.; Mislevy, R.; Bock, D. *Bilog-MG 3*; Scientific Software International, Inc.: Lincolnwood, IL , 2002.
11. Embretson, S. E.; Reise, S. P. *Item Response Theory for Psychologists*; Lawrence Erlbaum Associates, Inc.: Mahwah, NJ, 2000; p 371.
12. Baker, F. B. *The Basics of Item Response Theory*, 2nd ed.; ERIC Clearinghouse on Assessment and Evaluation: College Park, MD, 2001.

13. Rasch, G. *Probabilistic Models for Some Intelligence and Attainment Tests*; MESA Press: Chicago, 1960.
14. Rasch, G. On General Laws and the Meaning of Measurement in Psychology. In *Proceedings of the Fourth Berkeley Symposium on Mathematical Statistics and Probability*; University of California Press: Berkeley, 1961; pp 321–333.
15. Wilson, M. *Methods of Psychol. Res. Online* **2003**, *8*, 1–22.
16. The correlation data analysis for this paper was generated using SAS/STAT software, version 9.1.3 of the SAS System for Windows. Copyright 2004 SAS Institute, Inc. SAS and all other SAS Institute, Inc. Product or service names are registered trademarks or trademarks of SAS Institute, Inc., Cary, NC, U.S.A.
17. Caughran, J. A.; Martin, J. G.; Atwood, C. H. JExam: A Java-Based Exam Engine. In *Proceedings of the CONFCHEM*, October 13–19, 2001.
18. Atwood, C. H.; Martin, J. G.; Caughran, J. A. Internet-Based Testing in Freshman Chemistry. In *Proceedings of the 221st ACS National Meeting*; San Diego, CA, 2001.
19. Atwood, C. H.; Lautenschlager, G. J.; Marsh, R. L.; Martin, J. G.; Caughran, J. A. Initial Results from Computerized Assessment of Large Freshman Chemistry Classes at the University of Georgia. In *Proceedings of the 225th ACS National Meeting*, New Orleans, 2003, pp U417–U418.
20. Pinarbasi, T.; Canpolat, N. *J. Chem. Educ.* **2003**, *80*, 1328–1332.
21. Tien, L. T.; Teichert, M. A.; Rickey, D. *J. Chem. Educ.* **2007**, *84*, 175–181.
22. Ardac, D. *J. Chem. Educ.* **2002**, *79*, 510.
23. Birk, J. P.; Kurtz, M. J. *J. Chem. Educ.* **1999**, *76*, 124–128.
24. Tarhan, L.; Ayar-Kayali, H.; Urek, R. O.; Acar, B. *Res. Sci. Educ.* **2008**, *38*, 285–300.
25. Nicoll, G. *Int. J. Sci. Educ.* **2001**, *23*, 707–730.
26. Peterson, R. F.; Treagust, D. F. *J. Chem. Educ.* **1989**, *66*, 459–460.
27. Teichert, M. A.; Stacy, A. M. *J. Res. Sci. Teach.* **2002**, *39*, 464–496.
28. Khan, S. *Sci. Educ.* **2007**, *91*, 877–905.
29. Kruse, R. A.; Roehrig, G. H. *J. Chem. Educ.* **2005**, *82*, 1246–1250.
30. Harrison, A. G.; Treagust, D. F. *Sci. Educ.* **2000**, *84*, 352–381.
31. Teichert, M. A.; Tien, L. T.; Anthony, S.; Rickey, D. *Int. J. Sci. Educ.* **2008**, *30*, 1095–1114.
32. Chandrasegaran, A. L.; Treagust, D. F.; Mocerino, M. *Res. Sci. Educ.* **2008**, *38*, 237–248.
33. Nahum, T.; Mamlock-Naaman, R.; Hofstein, A.; Krajcik, J. *Sci. Educ.* **2007**, *91*, 579–603.
34. Bodner, G. M. *J. Chem. Educ.* **1991**, *68*, 385–388.
35. Nakhleh, M. B. *J. Chem. Educ.* **1992**, *69*, 191–196.
36. Robinson, W. R. *J. Chem. Educ.* **1998**, *75*, 1074.
37. Schmidt, H.-J. *Sci. Educ.* **1997**, *81*, 123–135.
38. Gabel, D. *J. Chem. Educ.* **1999**, *76*, 548–554.
39. Johnstone, A. H. *J. Comp. Assist. Learn.* **1991**, *7*, 701–703.
40. Devetak, I.; Urbancic, M.; Grm, K. S. W.; Krnel, D.; Glazer, S. A. *Acta Chim. Slov.* **2004**, *51*, 799–814.
41. Mulford, D. R.; Robinson, W. R. *J. Chem. Educ.* **2002**, *79*, 739–744.

42. Smith, K. J.; Metz, P. A. *J. Chem. Educ.* **1996**, *73*, 233–235.
43. Taber, K. S. *Int. J. Sci. Educ.* **1998**, *20*, 597–608.
44. Özmen, H. *J. Sci. Educ. Technol.* **2004**, *13*, 147–159.
45. Furió, C.; Calatayud, L. *J. Chem. Educ.* **1996**, *73*, 36–41.
46. Taagepera, M.; Arasasingham, R.; Potter, F.; Soroudi, A.; Lam, G. *J. Chem. Educ.* **2002**, *79*, 756–762.
47. Cokelez, A.; Dumon, A.; Taber, K. S. *Int. J. Sci. Educ.* **2008**, *30*, 807–836.
48. Ludwig, O. G. *J. Chem. Educ.* **2001**, *78*, 634.
49. Vaarik, A.; Taagepera, M.; Tamm, L. *J. Baltic Sci. Educ.* **2008**, *7*, 27–36.
50. Cervellati, R.; Perugini, D. *J. Chem. Educ.* **1981**, *58*, 568–569.
51. Ebenezer, J. V.; Erickson, G. L. *Sci. Educ.* **1996**, *80*, 181–201.
52. Raviolo, A. *J. Chem. Educ.* **2001**, *78*, 629–631.
53. Krieger, C. R. *J. Chem. Educ.* **1997**, *74*, 306–309.
54. Yin, M. *J. Chem. Educ.* **2001**, *78*, 1345–1347.
55. O'Farrell, F. J. Shared Meaning in Chemistry. In *Chemistry in Action*; 2000.
56. Gorin, G. *J. Chem. Educ.* **1994**, *71*, 114–116.
57. Greenbowe, T. J. *Int. J. Sci. Educ.* **2003**, *25*, 779–800.
58. Eubanks, I. D.; Eubanks, L. T. *Writing Tests and Interpreting Test Statistics: A Practical Guide*; American Chemical Society, Division of Chemical Education, Examinations Institute, Clemson University: Clemson, SC.

Chapter 11

All Assessment Materials Are Not Created Equal: The Myths about Instrument Development, Validity, and Reliability

Jack Barbera[*,1] **and Jessica R. VandenPlas**[2]

[1]**Department of Chemistry and Biochemistry, University of Northern Colorado, Greeley, Colorado 80639**
[2]**Department of Chemistry, Grand Valley State University, Allendale, Michigan 49401**
[*]**E-mail: jack.barbera@unco.edu.**

Educators and educational researchers strive to evaluate the impact of teaching on learning. To perform these assessments, we rely on instruments (e.g., surveys, questionnaires, tests, inventories) constructed of various items. But how do we know that the instrument itself produces valid and reliable results? Incorrect student responses on any assessment can stem from a number of sources. To truly uncover what information students understand, educators and researchers need to not only know whether students answers are correct or incorrect, they must uncover why they are correct or incorrect. Carefully constructed and validated assessment instruments can provide a useful tool for this purpose. This chapter will explore several myths regarding the development, validation, and reliability of assessment instruments. The goal is to both inform instrument users on what to look for when seeking out instruments as well as to inform instrument developers on various validation techniques.

The assessment of students is ubiquitous in chemistry education. We assess everything from student preparedness for a course to the long-term retention of knowledge. The results of these measures rely heavily on the quality of the assessment itself, specifically the instrument used during data collection. Educators and researchers use these data to not only assign grades to students, but to inform decisions on topics such as curricular impact, course reform, and knowledge structure.

Regardless of the focus of the assessment, the educator or researcher has two options: 1) develop an instrument or 2) use an existing instrument developed by others. This chapter is meant to inform both of these options. By providing details on key points of the instrument development process and clearly defining the variety of methods available to determine instrument data validity and reliability, the chapter can be used as a resource to aid in selecting or developing the proper assessment instrument for a particular task.

In this chapter we use the term "instrument" to refer to any ensemble of items (i.e., questions or statements) used to evaluate or probe characteristics of students. For example, an instrument constructed of multiple choice questions can be designed to gauge students' understanding of a specific chemistry concept, or an instrument can ask students a series of open-ended questions to gauge their self-efficacy about chemistry problem solving. An instrument designed to measure an educationally relevant characteristic of chemistry students should not be treated any differently than an instrument used in a chemical analysis. As chemists, we choose scientific instruments (e.g., HPLC, GC-MS, NMR) based on their intended purpose, accuracy, precision, and resolution. The quality of the data obtained from a scientific instrument depends on the quality of the instrument itself as well as how the instrument is used and calibrated. The same characteristics should be considered when choosing and using an educational instrument. Educational instruments are designed for a specific purpose (e.g., to measure student content knowledge of thermochemistry) and should be chosen based on their purpose. The term "validity" is used to describe the accuracy of the data derived from an educational instrument. This characteristic lets users determine if the instrument is measuring what it is suppose to measure (e.g., students' content knowledge of thermochemistry). The term "reliability" is used to describe the precision or reproducibility of the data derived from the instrument. This characteristic lets users determine the consistency of the data generated. These terms will be used throughout this chapter, which will discuss the properties of educational instruments.

Why should chemistry educators care about the quality of their assessment instruments?

The quality of an assessment instrument directly impacts the quality of the data that is derived from it. We need educational data to inform everything from student grades to the content of our courses. In order to make informed decisions on these or any other educational aspects of a course, instructors or researchers need to have an understanding of the quality of their data. For example, if an educator has designed a new methodology for teaching a concept, the educator needs to gather data to understand the impact of this methodology on student understanding of the target concept. If this data is not valid (accurate) nor reliable (precise), then the educator may be falsely promoting (or redesigning) the methodology. Decisions based on faulty data may impact the fairness of student grades, the adoption of new teaching tools/methodologies, or the content of courses.

This chapter is not meant to be a comprehensive training manual on instrument development, nor is it meant to provide an assessment or ranking of currently available instruments for use in chemistry education. This chapter's focus lies solely in addressing a few myths regarding assessment instruments.

Myth 1: An instrument can be validated.

An instrument itself cannot be shown to be valid or invalid. The data generated through the use of an instrument can be validated for a specific purpose with a specific population. This is a very easily overlooked fact when shopping for an assessment instrument. It is common to see journal articles referring to "valid" or "validated" instruments, these terms refer to the data generated, not the actual instrument. Several factors may lead to an instrument's data being rendered invalid.

1) If the instrument is used outside of the target population.
2) If the instrument is modified in any way, including:

    a) adding or subtracting items,
    b) changing the order of the items,
    c) altering scoring procedures,
    d) or modifying item language.

When an instrument is developed, the authors have a defined target population (e.g., students in an introduction to chemistry course). This population serves as the test sample on which the development, validity, and reliability results are based. Using the instrument with a population outside of this target population may alter the instrument's results. The data collected from this instrument can only be understood and interpreted in the context of this target population, as every individual holds preexisting knowledge that can affect how the individual interprets and responds to a question (*1*). Individuals outside of the target population (e.g., students in a majors' general chemistry course) may answer the same question in the same way, but for different reasons. These students may also provide answers based on context which is outside that of the target population, yielding inconclusive results from the instrument. Therefore, we cannot assume that an instrument developed for introductory chemistry students will provide equally valid results for chemistry majors in a general chemistry course.

Modifications to the instrument itself, which may seem miniscule or benign, may dramatically change what the instrument is measuring. This can render the resultant data invalid. The addition or subtraction of items from an instrument can affect the content validity (see Myth 3 section for discussion of content validity) or grouping of variables within the instruments subscales. Even the rearrangement of items can affect the measured outcome (*2*). While the delivery format (online or pencil-paper) of noncontent-based assessments does not affect their outcome (*3*), it may alter the response rate and sample population used to establish validity and/or reliability measures. Lastly, the way an instrument is scored may affect its results. For instance, Likert-scale responses may change based on the size of the scale (number of response options) (*4*), therefore, results may change if the scale is collapsed or expanded (*5*).

What if there is not an instrument that meets my needs?

Any instrument can be modified and its results revalidated under a different set of circumstances. This would expand the repertoire of useful assessment instruments for use in chemistry education.

Myth 1 (*An instrument can be validated*) Debunked:

Instruments cannot be validated; the data derived from them can be shown to be valid for a specific population under a specified set of circumstances.

Myth 2: Writing assessment items is easy.

In their development and validity protocol (*6*), Adams and Wieman state "the entire process [from assembly to evaluation] takes very roughly a half person year of a good Ph.D. level person's time to carry out". A large portion of this required time (at least 50%) goes into producing and validating the items themselves. In fact, the design and development of individual items for an instrument is the first step to ensuring an instrument produces valid and reliable results. The creation of instrument items has three distinct phases: 1) design, 2) development, and 3) validation.

Item design entails how the items are presented in the instrument (e.g., multiple choice, Likert-type, free response) and also refers to where the ideas or background information for the content of each item is derived. In terms of item presentation, in his book *Constructing Measures* (*7*), Wilson dedicates an entire chapter to the various formats. In lieu of being repetitive on this topic, interested readers are referred to that resource. The information that informs the content of the items themselves can come from either the author's experiences in working with the target population or reflect ideas from the literature. Either way, these ideas are used to define the span and depth of the instrument.

The development of individual items for an assessment instrument entails more than simply writing questions or statements. The development process cannot take place in a vacuum. That is to say that it is not acceptable in an assessment instrument for an author to write questions or statements without input and feedback that involves the target population. Although it is acceptable to use one's own experience in designing the general ideas for the items, items developed without input from representatives of the target population often are misinterpreted due to ambiguity in meaning or unclear vocabulary. For example, a simple statement such as "I find equations helpful in understanding chemistry" can have multiple problems. First, does the word "equation" refer to chemical equations or mathematical equations? Second, what does it mean to be "helpful" or to "understand" chemistry? The author of the item may have very specific ideas of what information he/she will get from the item regarding student learning. However, if students in the target population misinterpret the words or have very different meanings for its phrases then what is the item actually measuring? It is difficult for educators and researchers to remember back to freshman chemistry and try to "think like a student". Valuable information (and maybe even new ideas) about instrument items can be derived from interviews with students. Issues concerning an item's wording and interpretation are quickly revealed in an interview setting where students are asked to explain what the question is asking (*7*). Such information regarding their interpretation of question wording, graphs, figures, or equations used in an item can help the author develop a clear and concise question.

Finally, before the results of an instrument as a whole can be validated, each item must individually be shown to produce valid results with the target population. This "item-level" validity begins with ensuring clarity in wording and meaning during development, as discussed above. Once items are developed, a pilot version of the instrument can be assembled. The pilot instrument should be administered to the target population under the same conditions (e.g., one hour in-class with bubble sheets) intended for the final version of the instrument. Results for each item can then be evaluated for characteristics such as difficulty and discrimination, these properties will aid in identifying items that do not function as expected. For example, data may show that an item expected to be easy is actually very difficult for the target population. This result may be due to a problem with the wording or a representation used within the item; in any case, the item would need further attention.

Myth 2 (*Writing assessment items is easy*) Debunked:

The validity of an assessment instrument's data begins with the design and development of its items. This process entails more than simply writing a set of questions.

Myth 3: Validity is a purely subjective measure.

Once items are assembled into an instrument, data can be collected to gauge the validity and reliability of the instrument in a given situation. Validity can be thought of as the accuracy of the instrument (*8*), that is, does the instrument measure what it is purported to measure? Validity can be measured in a variety of ways, each providing a unique set of data about the instrument. There are two main types of validity, construct validity and criterion validity, each with their respective subsets. **Construct validity** refers to whether an instrument's measure matches or correlates to the conceptual (or theoretical) variable it is designed to measure. It is construct validity that answers the question: "Does the instrument actually measure what we think it is measuring?" For example, an instrument designed to gauge a student's understanding of chemical kinetics should produce data that supports a student's knowledge (or lack of knowledge) of this topic. This differs from criterion validity in that **criterion validity** (sometimes called concrete validity) is the extent to which an instrument's measures are related to some concrete criteria in the "real world". Criterion validity answers the question: "Does the instrument predict the outcome of some other measure?" For example, in the chemical kinetics example above, does performance on the instrument correlate to other measures of performance (e.g., a kinetics experiment, problem set, or stock exam questions)? Both construct and criterion validity can be broken down into several subcategories, outlined in Figure 1. It is important to note that not every subcategory needs to be accounted for to establish the validity of an instrument's results. The type of instrument and the intended use of its results determine the required validity measures. For example, an instrument designed as an assessment of students' mathematical ability at the start of a physical chemistry course would be expected to show predictive validity in terms of students' course grade. However, an assessment of the same students' beliefs about the nature of science would not be expected to have predictive validity in terms of their course grade.

Construct validity is the most widely reported form of instrument validity. Each of the subcategories shown in Figure 1 establishes a different form of construct validity and hence means something different about the instrument itself.

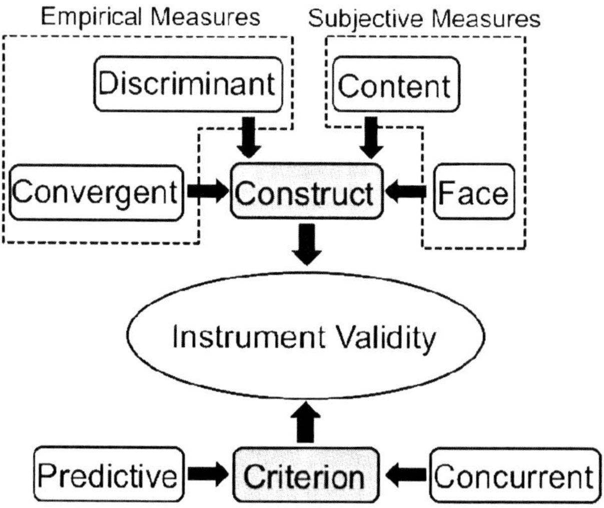

Figure 1. Map showing the main categories of Construct and Criterion validity along with their respective subcategories. Each subcategory evaluates a unique aspect of instrument validity.

Face validity is a subjective measure of construct validity that simply refers to whether the instrument *appears* to measure a proposed variable. To establish face validity, an instrument developer needs to do no more than ask someone else to look over the instrument and comment on if they "think" it will measure what the author purports it to measure (e.g., knowledge of chemical bonding). This is the weakest method for demonstrating construct validity. In the social and behavior sciences, it is believed that this can be established using either novices or experts as evaluators. However, for content related instruments, face validity should be established with knowledgeable experts. It would be unlikely that a novice (i.e., non-chemist) could provide face validity on a chemistry concept instrument or any other chemistry-content-rich assessment.

Content validity (sometimes called logical validity) refers to the coverage of the content domain, as defined by the instrument developer, as well as the importance and utility of the instrument. This subjective measure is established by having a series of content experts review the instrument. A widely used measure to establish content validity is to gauge the agreement between experts on the importance or utility of an instrument's items. For example, to establish content validity for an instrument on general chemistry student misconceptions of thermochemistry, an instrument developer would recruit chemists who regularly teach this course. These content experts would comment on the coverage of thermochemistry topics and the proposed misconceptions being addressed. In his publication titled *A quantitative approach to content validity* (*9*), Lawshe established a formula to calculate the content validity on an item that accounts for the agreement and disagreement between expert evaluators.

While face and content validity should be established for all developed instruments, they are relatively subjective measures and should not be the end of the validation studies. Using one of the empirical measures (Figure 1) can strengthen the construct validity of an instrument.

Convergent validity and **Discriminant validity** can be thought of as contrasting properties that relate to the correlation between the variable measured by the instrument and some other measured variable. Convergent validity would be established by showing that measures that *should be related* are in fact related (positively correlated). For example, the convergent validity of an instrument designed to measure a student's chemistry content knowledge can be established by comparing the instrument's results to some other stock assessment (e.g., an American Chemical Society Exam (*10*)). One would expect these two measures to show a positive correlation. In contrast, discriminant validity would be established by showing that measures that *should not be related* are in fact not related (uncorrelated). For example, the results from an instrument designed to measure a student's molecular level understanding of chemistry should not be related to the student's mathematical or reading abilities. These measures should show little to no correlation. Discriminant validity can be used to ensure that there are no confounding factors to the instrument, such that you can say the instrument's results are not dependent on reading ability or mathematical ability, but rather measure only the specific variable intended.

The subcategories of criterion validity are also shown in Figure 1. One type of criterion validity, called **predictive validity,** relates to how well an instrument predicts an other externally measured variable. In a study of predictive validity, the instrument scores are collected first; then at some later time the external variable is measured. A correlation between the two measures can then be established and the instrument can be evaluated on its ability to predict the additional external variable. For example, the predictive validity of standardized college entrance exams, such as the SAT or ACT, has been established by correlating test scores with college performance. In most cases, these exams are highly predictive of student performance in the measured areas.

The **concurrent validity** of an instrument is its ability to measure something it should theoretically be able to measure. For example, it is expected that an instrument designed to measure students' chemistry content knowledge should be able to discriminate between introductory chemistry students and graduating chemistry majors. Therefore, when this instrument is administered to these two theoretically different groups of students, it should produce results showing that the graduating students (hopefully) have more chemistry content knowledge than the introductory chemistry students.

Myth 3 (*Validity is a purely subjective measure*) Debunked:

The validity of the data generated from any instrument entails both subjective and empirical data, which when combined provide a basis for the accuracy of the instrument.

Myth 4: Reliability implies validity.

The reliability of an instrument is distinctly different from the validity of the instrument's data. Validity is a measure of the *accuracy* of the data; reliability is an assessment of the *random error* present in an instrument's data. Reliability of the data does not imply validity, an instrument can be relatively free of random error thereby producing consistent data, but may not be an accurate measure of the expected variable. An instrument needs to be evaluated for both components using data generated with the target population.

The reliability of an instrument can be estimated in two different categories, internal consistency and reproducibility. The **internal consistency** of an instrument establishes the correlation among its items and evaluates whether all items are measuring the true score rather than producing random error. The **reproducibility** of an instrument is an estimate of how item scores correlate to one another during measurements taken at different times. Internal consistency measurements only require a single administration of the instrument, as opposed to reproducibility measurements that require two or more administrations.

The internal consistency of an assessment instrument using scaled items can be measured in a variety of ways. The two most widely used methods in establishing internal consistency are the split-half method and the Cronbach's alpha calculation. **Spilt-half reliability** is determined by randomly dividing the instrument items into two sets. The correlation coefficient between the scores of the two sets can then be used as an estimate of the consistency among items. Because this method specifically uses two predetermined sets of data, it does not give the most robust evaluation of the instrument's internal consistency. A more thorough analysis would be to perform the split-half method with every combination of sets for an instrument. **Cronbach's alpha** is essentially the mathematical average of all possible split-half measurements. The alpha coefficient is easily calculated by most statistical programs and is the most widely reported reliability measure. Cronbach's alpha scores range from 0.00 (complete random error) to 1.00 (no random error). Ideally, the alpha value should be greater than 0.7 (*8*) to establish good internal consistency, however, too high an alpha value (>0.9) may imply significant redundancy within the instrument items (*11*). The Kudner-Richardson (KR-20) coefficient is comparable to Cronbach's alpha but used for dichotomous data.

The reproducibility of an instrument is measured using the **test-retest method**. This reliability measure requires two separate administrations of the same instrument to the same group of participants. The correlation coefficient between results is an estimate of the instrument's reproducibility. Retesting effects may confound this method. Mainly, the time between administrations may alter a participant's responses. If the time is *too short*, the results may be affected by memorization. For instance, if participants know how they responded during the first administration, they may be likely to answer differently if they believe that that is what the researcher wants (why else would the researcher give the same questions twice?). Or, respondents may try to simply duplicate their first answers so as not to seem inconsistent. If the time is *too long*, the results of an individual may be affected by true changes in the variable being measured. For instance, an instrument designed to measure students' expectations about a course will be highly influenced by the course itself. Therefore, if the administrations are spaced too far apart they may be measuring changes in expectations over time, yielding an ambiguous reproducibility score. When reporting test-retest data, an instrument designer should provide details on the time frame as well as how these potential confounding factors were addressed.

The last assessment issue that falls under the reliability umbrella is that of inter-rater reliability. **Inter-rater reliability** is a measure of the consistency among a group of evaluators on non-scale items. This measure becomes important whenever a participant's response needs to be interpreted to yield a score or result. Instruments that contain free-response items asking participants to "explain" their answer are usually scored using a rubric. The reliability of the data therefore lies in the consistency of interpretation. The consistency between evaluators is calculated using the kappa statistic, which ranges from 0 (all random error) to 1 (no random error) and is easily calculated using most statistical software packages. Values of kappa greater than 0.8 are "very good", values between 0.6-0.8 are "good", those below 0.6 are "moderate to poor". If the scores from independent raters are below the "good" range then the raters need to be trained or the rubric modified to produce more consistent results (*12*).

Myth 4 (*Reliability implies validity*) Debunked:

Instrument reliability is a complementary measure to the validity of its data. An instrument's data must be shown to be valid (accurate) first, and then reliability should be determined. A measure such as a Cronbach's alpha or KR-20 does not imply that the instrument produces valid results, only that it has internal consistency.

Myth 5: All assessment instruments are created equally.

There are many existing assessment instruments published in peer-reviewed journals and readily available on non-journal web sites. For the purpose of this chapter, these instruments fall into two broadly defined categories, content and noncontent-based. Content-based assessments are those that target student knowledge about chemistry or chemistry related material. Noncontent-based assessments are those that address areas such as student attitudes, goals, or motivation.

Regardless of the category, there is no universal methodology for the development protocol of an assessment instrument. Tables 1 through 4 present an overview of some currently available instruments being used in chemistry education. This is not an exhaustive list, the examples provided cover a range of topics, development procedures, and validity/reliability studies. Chemistry educators and researchers looking for an instrument to use in their classrooms or research are encouraged to utilize the vast array of chemistry and science education journals available to uncover many more instruments than can be presented here. The tables are meant to show the broad range of procedures utilized by instrument developers for their respective instruments. These tables provide a quick reference list of each instrument's general details (Tables 1 and 3) and an overview of their development procedures and validity/reliability studies (Tables 2 and 4). These tables are **NOT** a comprehensive analysis of the instruments, and do not provide any determination on the quality of the instruments themselves. These tables comprise information provided by developers in journal articles and conference proceeding regarding their readily available instrument. Readers are encouraged to consult each instruments reference for further details and definitions. As each category in the tables encompasses a vast amount of information, each table should be used as a guide only. Previous sections in this chapter explain many of the tables' variables.

Myth 5 (*All assessment instruments are created equally*) Debunked:

There is a wide range of development protocols being used for chemistry education instruments; these may differ based on the type of instrument and its intended goal. Some protocols may provide more robust instruments than others.

Table 1. General details of the content-based instruments used in chemistry education

Title	*Topic*	*Number of Items*	*Response Type*	*Target Population*
Covalent Bonding and Structure Concepts **(CBSC)** (*13*)	Covalent Bonding and Structure	15	Two-Tier Multiple Choice	High School[a] Chemistry
Chemistry Concept Inventory **(CCI)** (*14*)	First Semester General Chemistry	22	Multiple Choice	College Chemistry
Chemistry Concept Inventory **(ChCI)**[b] (*15*)	First and Second Semester General Chemistry	20 (each)	Multiple Choice	College Chemistry
Introductory Thermal Concept Inventory **(ITCE)** (*16*)	Thermodynamics	26	Multiple Choice	High School/College[c] Chemistry and Physics
Mole Concept Test (MCT) (*17*)	Mole Concept	20	Varies based on question[d]	College Chemistry
Conceptions of the Particulate Nature of Matter and Understanding of Chemical Bonding **(PNCB)** (*18*)	Particulate Nature of Matter and Chemical Bonding	10	Two-Tier Multiple Choice	High School[e] Chemistry
Test to Identify Student Conceptualization **(TISC)** (*19*)	Chemical Equilibrium	10	Two-Tier Multiple Choice	College Chemistry

[a] Test population: Grade 11 – 12 Australian students. [b] Two separate instruments, same development protocols, different study populations. [c] Test population: Grade 10 – 13 Australian students. [d] Multiple types of questions:

1) Multiple-Choice
2) Two-Tier True-False
3) Two-Tier Multiple-Choice
4) Problems

[e] Test population: Grade 9 and 10, boys school in Singapore.

Table 2. Development and validity/reliability details of the content-based instruments used in chemistry education. Mark (X) denotes information reported by instrument developers.

			CBSC	*CCI*	*ChCI*[a]	*ITCE*	*MCT*	*PNBC*	*TISC*
Question/Statement Development	Literature Basis			X	X	X	X	X	X
	Research with Target Population	Interviews	X		X	X			X
		Pilot Study	X	X	X				
	Author Derived		X	X	X	X	X	X	X
Item Statistics	Difficulty		X		X	X	X	X	
	Discrimination		X		X	X	X	X	
Validity Studies	Construct	Face	X	X	X	X	X		X
		Content	X	X	X		X	X	X
		Discriminant							
		Convergent			X				X
	Criterion	Concurrent				X			
		Predictive							
Reliability Studies	Internal Consistency		X	X	X	X	X	X	X
	Test – Retest								

[a] Two separate instruments, same development protocols, and different study populations.

Table 3. General details of the noncontent-based instruments used in chemistry education

Title	*Topic*	*Number of Items*	*Response Type*	*Target Population*
Attitude toward the Subject of Chemistry Inventory **(ASCI)** *(20)*	Attitudes	20 (5 categories)	Semantic Differential	College Chemistry (General)
Chemistry Attitudes and Experiences Questionnaire **(CAEQ)** *(21)*	Attitudes, Self-Efficacy, and Learning Experiences	21/17/31[a]	Semantic Differential/Likert[b]	College Chemistry (General)
Chemistry Expectations Survey **(CHEMX)** *(22)*	Expectations	47 (7 categories)	Likert	College Chemistry (Community and General)
Colorado Learning Attitudes about Science Survey **(CLASS)** *(23)*	Beliefs	50 (9 categories)	Likert	College Chemistry (Nonscience through Graduate)
Chemistry Self-Concept Inventory **(CSCI)** *(24)*	Self-Concept	40 (5 categories)	Likert	College Chemistry (Nonscience and General)
Group Assessment of Logical Thinking **(GALT)** *(25)*	Piagetian Logical Thinking Ability	21[c]	Multiple Choice	Grade 6 - College
Metacognitive Activities Inventory **(MCAI)** *(26)*	Metacognition in Chemistry Problem Solving	27	Likert	College Chemistry (General through Graduate)
Motivated Strategies for Learning Questionnaire **(MSLQ)** *(27)*	Motivation and use of Learning Strategies	81 (2 sections/15 subscales)	Likert	College students in any discipline

Continued on next page.

Table 3. (Continued). General details of the noncontent-based instruments used in chemistry education

Title	*Topic*	*Number of Items*	*Response Type*	*Target Population*
Purdue Visualization of Rotations Test **(ROT)** (*28*)	Spatial Ability	20	Multiple Choice	College Chemistry (General and Organic)
Student Assessment of Learning Gains **(SALG)** (*29*)	Course Evaluation	46[d]	Likert	College Chemistry (General)

[a] Number depends on section (attitude/self-efficacy/learning experiences). [b] Attitude and self-efficacy sections are semantic differential, learning experiences is Likert. [c] Original instrument contains 21 items, a 12 item version does exist but validation data could not be found. [d] There are 46 items present in the "Template" version of the SALG. Users are encouraged to add, delete, or modify these questions to fit their particular needs.

Table 4. Development and validity/reliability details of the noncontent-based instruments used in chemistry education. Mark (X) denotes information reported by instrument developers.

			ASCI	*CAEQ*	*CHEMX*	*CLASS*	*CSCI*	*GALT*	*MCAI*	*MSLQ*	*ROT*	*SALG*[d]
Question/Statement Development	Literature Basis		X	X	X	X	X	X		X		
	Research with Target Population	Interviews		X		X		X				X
		Pilot Study		X			X	X	X	X		
	Author Derived		X	X	X	X			X	X	X	X
Validity Studies	Construct	Face		X		X		X	X			
		Content	X	X			X		X	X		
		Discriminant	X							X	X[c]	
		Convergent					X	X	X	X	X[c]	
	Criterion	Concurrent	X	X	X	X	X		X		X	
		Predictive		X				X		X		
Reliability Studies	Internal Consistency		X	X	X[a]	X	X	X	X	X	X	
	Test – Retest		X			[b]	X		X			

[a] Based on faculty responses. [b] Authors report Test-Retest data but this reliability measure was incorrectly performed. [c] Based on original 30 item instrument, not the currently available 20 item version. [d] Authors state that the original 46 items could not be shown valid or reliable, as users are encouraged to add, delete, or modify items to fit their particular needs, no validity or reliability data exists.

References

1. Bransford, J. D.; Brown, A. L.; Cocking, R. R.; Donovan, M. S.; Pellegrino, J. W. *How People Learn: Brain, Mind, Experience, and School*; National Academy Press: Washington, DC, 2000.
2. Schuman, H.; Presser, S. *Questions and Answers in Attitude Surveys: Experiments on Question Form, Wording, and Context*; Academic Press: New York, 1981.
3. Denscombe, M. *Soc. Sci. Comput. Rev.* **2006**, *24*, 246–254.
4. Crocker, L.; Algina, J. *Introduction to Classical and Modern Test Theory*; Holt, Rinehart, and Winston: Fort Worth, TX, 1986.
5. McCall, C. H. An Empirical Examination of the Likert Scale: Some Assumptions, Development, and Cautions. http://faculty.pepperdine.edu/cmccall/CERAFinal.pdf.
6. Adams, W. K.; Wieman, C. E. *Int. J. Sci. Educ.* **2010**, DOI:10.1080/09500693.2010.512369, pp 1−24.
7. Wilson, M. *Constructing Measures: An Item Response Modeling Approach*; Lawrence Erlbaum Associates: Mahwah, NJ, 2005.
8. Stangor, C. *Research Methods for the Behavioral Sciences*, 4th ed.; Wadsworth, Cengage Learning: Belmont, CA, 2010.
9. Lawshe, C. H. *Pers. Psychol.* **1975**, *28*, 563–575.
10. American Chemical Society, Division of Chemical Education, Examinations Institute. http://chemexams.chem.iastate.edu/ (accessed June 2011).
11. Streiner, D. L.; Norman, G. R. *Health Measurement Scales: A Practical Guide to Their Development and Use*; Oxford Univeristy Press: New York, 1989.
12. Landis, J. R.; Koch, G. G. *Biometrics* **1977**, *33*, 159–174.
13. Peterson, R.; Treagust, D.; Garnett, P. *Res. Sci. Educ.* **1986**, *16*, 40–48.
14. Mulford, D. R.; Robinson, W. R. *J. Chem. Educ.* **2002**, *79*, 739–744.
15. Krause, S.; Birk, J.; Bauer, R.; Jenkins, B.; Pavelich, M. J. *Development, Testing, and Application of a Chemistry Concept Inventory*, ASEE/IEEE Frontiers in Education Conference, Savannah, GA, 2004.
16. Yeo, S.; Zadnik, M. *Phys. Teach.* **2001**, *39*, 496–504.
17. Krishnan, S. R.; Howe, A. C. *J. Chem. Educ.* **1994**, *71*, 653–655.
18. Othman, J.; Treagust, D. F.; Chandrasegaran, A. L. *Int. J. Sci. Educ.* **2008**, *30*, 1531–1550.
19. Voska, K. W.; Heikkinen, H. W. *J. Res. Sci. Teach.* **2000**, *37*, 160–176.
20. Bauer, C. F. *J. Chem. Educ.* **2008**, *85*, 1440–1445.
21. Dalgety, J.; Coll, R. K.; Jones, A. *J. Res. Sci. Teach.* **2003**, *40*, 649–668.
22. Grove, N.; Bretz, S. L. *J. Chem. Educ.* **2007**, *84*, 1524–1529.
23. Barbera, J.; Adams, W. K.; Wieman, C. E.; Perkins, K. K. *J. Chem. Educ.* **2008**, *85*, 1435–1439.
24. Bauer, C. F. *J. Chem. Educ.* **2005**, *82*, 1864–1870.
25. Bitner-Corvin, B. *Is the GALT a Reliable Instrument for Measuring the Logical Thinking Abilities of Students in Grades Six through Twelve?* National Association for Research in Science Teaching Conference, Lake Ozark, MO, 1998.

26. Cooper, M. M.; Sandi-Urena, S. *J. Chem. Educ.* **2009**, *86*, 240–245.
27. Pintrich, P. R.; Smith, D. A. F.; Garcia, T.; Mckeachie, W. J. *Educ. Psychol. Meas.* **1993**, *53*, 801–813.
28. Bodner, G. M.; Guay, R. B. *Chem. Educ.* **1997**, *2*, 1–17.
29. Seymour, E.; Wiese, D. J.; Hunter, A. B.; Daffinrud, S. M. *Creating a Better Mousetrap: On-Line Students Assessment of Their Learning Gains*, National Meeting of the American Chemical Society, San Francisco, CA, 2000.

Chapter 12

Assessing Conceptual versus Algorithmic Knowledge: Are We Engendering New Myths in Chemical Education?

Thomas Holme*[,1] and Kristen Murphy[2]

[1]Department of Chemistry, Iowa State University, Ames, Iowa 50011
[2]Department of Chemistry and Biochemistry, University of Wisconsin – Milwaukee, Milwaukee, Wisconsin 53201
***E-mail: taholme@iastate.edu**

Studies over the past two decades have emphasized a gap between relatively weak student performance on conceptual items versus traditional items. The ACS Examinations Institute has released a pair of exams for general chemistry in which items are intentionally paired with one conceptual and one traditional item. This paper describes data from statistical analysis of the item pairs, and notes that for these exams, this gap is not evident, as overall performance is better on conceptual items. Possible implications for teaching and for research in Chemistry Education are noted.

Introduction

The prospect that students may learn quantitative problem solving skills within chemistry while not understanding the conceptual basis for the content has been of interest for over 20 years. For example, Nurrenburn and Pickering found that conceptual understanding of stoichiometry lagged behind quantitative understanding (*1*). Subsequently, several groups have confirmed this as well as determined other features. Pickering established (*2*) that performance on conceptual questions in general chemistry was not a predictor of success in organic chemistry. Sawrey showed that difficulties with conceptual items were found for students with both high and low performance on traditional quantitative items (*3*). Nakhleh and coworkers carried out a series of studies that further established the gap between conceptual understandings and algorithmic problem solving skills and sought pedagogies to mediate that gap (*4–8*). More recent work by Niaz (*9*) and Cracolice (*10*) continues to identify ways in which conceptual knowledge lags behind algorithmic knowledge in chemistry students. A key component of all of these studies was the use of the paired-question format, where student performance comparisons are drawn from multiple choice item pairs that are designed to provide data about conceptual and algorithmic knowledge separately. The ACS Exams Institute provided a specific tool for this type of assessment in 1997 (*11*) and updated the general chemistry paired questions exams in 2005 and 2007 (*12*, *13*).

The importance of conceptual misunderstandings that were uncovered via this methodology led to a wide range of studies that identified student misconceptions (or alternate conceptions) in a number of content domains of chemistry (*14–16*). In addition to identifying the existence of misconceptions, it is arguable that this data led to changes in the manner in which textbooks presented information about chemistry at the particulate level. Thus, over the past 20 years, since the conceptual/algorithmic gap was first uncovered, there has been both further research and pedagogical responses.

This chapter provides information about student performances on the 2005 Paired-Questions First-Semester General Chemistry Exam (GC05PQF) and the 2007 Paired-Questions Second-Semester General Exam (GC07PQS) that were released by the ACS Exams Institute. These exams have been used nationally for several semesters, and the norm generation process of the Institute (*17*) has allowed for the consideration of item-level analysis of each exam over several thousand student performances. This information can be used to establish two key things. First, the previously identified gap that was abundantly clear 20 years ago (*1–3*) is perhaps not as prominent or as unidirectional today. There are currently item pairs for which student performance on the conceptual item is better than on the algorithmic item. Second, these paired-questions exams show how important the measurement of student performance can be in terms of understanding what students are learning in general chemistry. In particular, exam design plays a critical role in observations about student learning. The current exams show a smaller gap, not only because the student performance database is sensitive to changes in pedagogy and curriculum over the past 20 years, but also because the designers of the exam itself sought an instrument in which the conceptual items and traditional items had more nearly equal performances. This chapter provides abbreviated national normative data for the two exams and also considers guidelines for how the results of these particular assessments are best framed in terms of current pedagogy and learning, and measurement theories.

Exam Development and Structure

The paired questions exams were prepared in a manner similar to the standard procedure for ACS Exams (*17*). The key difference is that not all items in each exam were developed originally for the exam. A number of items were obtained from already released exams. Nonetheless, after all workable item pairs were gleaned from available items on released exams, it was determined that some content areas were not adequately covered and specific items or item pairs were developed for the exam. A trial-test phase of the development was undertaken so that student performances could provide statistical data to determine which pairs of items to include on the released exam. This process also led to the development of pairs of items that illustrate item pairs in general chemistry, while not having the security restrictions that forbid the publication of items from ACS Exams. One such pair, from the first-term exam is shown in Figure 1.

C1. As shown in Figure 1, a gas is sealed into a syringe and immersed in a liquid at room temperature (300 K). When the liquid boils the gas in the syringe changes volume as depicted in Figure 2. Assuming no gas leakage and no change in atmospheric pressure, which is the best estimate of the boiling temperature of the liquid?

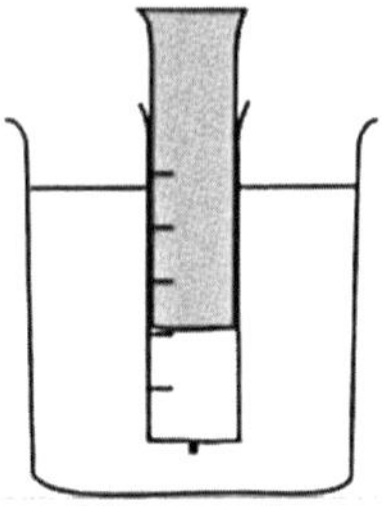

Figure 1

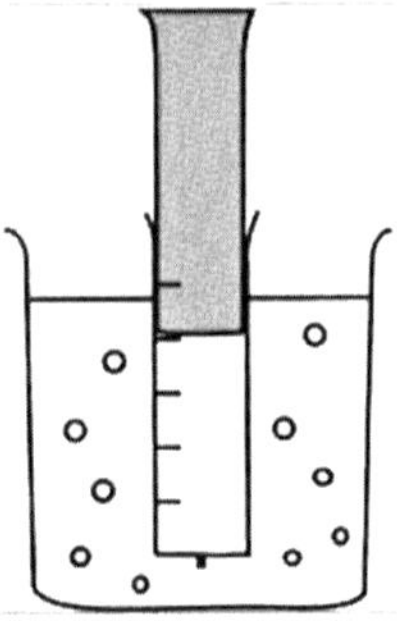

Figure 2

(A) about 200 K (B) about 300 K

(C) about 400 K (D) about 600 K

T1. A sample of CO_2 occupies 6.00 L at 25 °C and 700. mmHg. What is the volume of the gas at STP?

(A) 4.17 L (B) 5.06 L (C) 6.03 L (D) 22.4 L

Figure 1. Illustration of paired questions. The content area for this pair is gas laws. C1 is classified as conceptual, while T1 is classified as traditional.

Note that this pair was not used on the released exam so it can be reproduced here and referred to as an example of the paired-item format. It is also important to note that this conceptual item does not involve particulate-level representations of chemical systems as were used in the original work (*1–8*). There are conceptual items on the exam that utilize diagrammatic representations, but the construction of conceptual items is notably broader than this construct. Diagrammatic representations include graphical representations of data, such as a phase diagram or the image included in Figure 1, but do not include schematic depictions of matter at the particulate level.

Given this basic structure for item pairs, the overall released exams are constructed from 20 pairs. Tables I and II provide the overall structure of the exams in terms of content.

Table I. Content coverage of item pairs for First Term(*18*)

Topic	*Number of item pairs*
Properties of Matter	3 (6 items)
Atoms, Elements and Compounds	1 (2 items)
Gases	3 (6 items)
Stoichiometry	5 (10 items)
Solutions and Concentration	2 (4 items)
Atomic Structure	1 (2 items)
Molecular Structure	4 (8 items)
Thermochemistry	1 (2 items)

Table II. Content Coverage of item pairs for Second Term(*18*)

Topic	*Number of item pairs*
Equilibrium	3 (6 items)
Kinetics	2 (4 items)
Thermodynamics	3 (6 items)
Electrochemistry	3 (6 items)
Solutions	3 (6 items)
Acid/Base chemistry	4 (8 items)
Nuclear chemistry	2 (4 items)

Data that is returned for norm purposes, and reported here, is from students who were allowed 55 minutes (maximum) to complete the released exam. Instructors who purchase the exam are provided with the specific pairings. The conceptual item occurs earlier in the exam than the traditional item, in 39 of the 40 pairs of items. (The exception occurs in kinetics for the second-term exam, and was the result of formatting issues related to the items and their locations on the page.)

Basic Item Statistics

The Exams Institute calculates item statistics for all released exams based on classical test theory. The item difficulty is assigned as the fraction of students who answer it correctly. This definition results in a counter-intuitive scale, where an items with a higher difficulty value is answered correctly by more students. The second commonly calculated item statistic is the discrimination, calculated by subtracting the fraction of correct answers among the bottom performing students (as determined by their total score on the exam) from the fraction correct among the top performing students. The number of students in the sample for "top" and "bottom" may be varied, and for the values presented in this work the top quarter and bottom quarter of students are used in the calculations. Additional information about overall norms and item statistics have been published elsewhere (*18*).

The item statistics presented for the first-term exam are determined from 3073 student performances from 12 colleges who contributed data voluntarily. For the second-term exam the sample is derived from 3557 students from 9 colleges. Some large courses are included in this data, and in these cases, there are typically multiple sections present, thus there are more than 21 instructors associated with the data included here. Schools that returned data included large research institutions, comprehensive universities, liberal arts colleges and community colleges. The majority of student performances come from large research institutions in part because of the large size of the general chemistry classes at these institutions.

Looking at the data in Tables III and IV there are several key points to consider. First, the number of items for which student performance is better on the conceptual item than the traditional is 10/20 for the first term and 9/20 for the second term exam. Second, the average difficulty for conceptual items is 0.653 and 0.513 respectively for the first term exam and second term exam. The average difficulty for traditional items is 0.598 and 0.538 respectively. Thus, in terms of a mean behavior, conceptual item performance is slightly better on the first term exam and traditional item performance is slightly better on the second term exam. Finally, in topics where there are more than one item-pair it is quite uncommon, over the entire content domain, for performances to exclusively favor one style of item or another. The only examples are in second term thermodynamics and nuclear chemistry where the traditional item shows higher performance in all pairs for that content.

These observations are clearly tied to the test design. Because item choices for the released exams are predicated on performances based on the trial test phase of the exam, the exam committee made conscious choices to have pairs that favor one type of item over another. For example, at the trial exam stage for the first term exam, the average difficulty and discrimination of the conceptual items that were chosen for the released GC05PQF exam were 0.650 and 0.408 respectively. The traditional items selected for the released exam had 0.625 for difficulty and 0.483 for discrimination. Thus, the design of the exam was to have the conceptual and traditional items similar in difficulty. The "predicted" difficulty was quite close for the conceptual items, but the traditional items have now tested slightly more difficult than when they were trial tested. This was also true in the second term exam, where the predicted average difficulty based on the trial tests was 0.53 for conceptual items and 0.63 for traditional items (as compared to the observed 0.51 and 0.54.)

Finally, it is also worth noting that the two exams have some apparent structural differences as delineated in Table V. The difference in performance between the conceptual and traditional by item pairs is examined further, identifying a performance gap (by difficulty value) of more than 10% or 20%. It should also be noted that the item pairs included in the >20% category are also explicitly listed in the >10% category.

Table III. Classical Item Analysis for Paired Questions – First Term† (*18*)

Topic	*Item Pair*	*Conc. Diff.*	*Conc. Disc.*	*Trad. Diff.*	*Trad. Disc.*
Properties of Matter	P1	0.604	0.476	0.712	0.414
	P2	0.870	0.222	0.795	0.419
	P3	0.651	0.489	0.738	0.417
Atoms	A1	0.600	0.559	0.696	0.493
Stoichiometry	ST1	0.519	0.524	0.764	0.454
	ST2	0.812	0.334	0.419	0.547
	ST3	0.851	0.334	0.696	0.455
	ST4	0.456	0.551	0.607	0.680
	ST5	0.460	0.463	0.473	0.715
Gases	G1	0.655	0.479	0.698	0.39
	G2	0.741	0.390	0.715	0.454
	G3	0.609	0.434	0.188	0.338
Solutions	SO1	0.613	0.482	0.636	0.244
	SO2	0.445	0.421	0.404	0.547
Atomic Structure	AS1	0.557	0.423	0.611	0.385
Molecular Structure	MS1	0.611	0.433	0.687	0.490
	MS2	0.611	0.441	0.447	0.562
	MS3	0.866	0.260	0.651	0.488
	MS4	0.804	0.399	0.504	0.325
Thermo-chemistry	TH1	0.729	0.432	0.514	0.473

† Conc. Diff. = difficulty of conceptual item; Conc. Disc. = discrimination of conceptual item; Trad. Diff. = difficulty of traditional item; Trad. Disc. = discrimination of traditional item.

In the first-term exam, it is much more common for large differences (defined somewhat arbitrarily as greater than a 10% difference in performance) in difficulty for paired items to result from the conceptual item having a much higher difficulty index (i.e. the conceptual item has better performance.) By contrast, in the second-term exam, while the number of pairs with 20% or more performance difference is the same (2 each) the number of traditional items with at least 10% better perfomance is 6 compared to just 2 where the conceptual item shows better performance.

Table IV. Classical Item Analysis for Paired Questions – Second Term†

Topic	*Item Pair*	*Conc. Diff.*	*Conc. Disc*	*Trad. Diff.*	*Trad. Diff.*
	EQ1	0.486	0.442	0.532	0.368
Equilibrium	EQ2	0.495	0.583	0.408	0.486
	EQ3	0.224	0.391	0.445	0.452
Kinetics	K1	0.489	0.433	0.480	0.372
	K2	0.665	0.251	0.622	0.477
	TD1	0.445	0.375	0.561	0.337
Thermodynamics	TD2	0.426	0.427	0.707	0.441
	TD3	0.520	0.429	0.574	0.511
	EC1	0.403	0.435	0.333	0.382
Electrochemistry	EC2	0.513	0.441	0.613	0.475
	EC3	0.576	0.416	0.606	0.470
	SO1	0.462	0.394	0.417	0.417
Solutions	SO2	0.539	0.273	0.453	0.515
	SO3	0.511	0.485	0.749	0.501
	AB1	0.512	0.629	0.629	0.512
	AB2	0.766	0.412	0.422	0.330
Acids/Bases	AB3	0.608	0.449	0.599	0.386
	AB4	0.535	0.381	0.289	0.411
Nuclear	N1	0.561	0.469	0.611	0.443
	N2	0.528	0.475	0.700	0.497

† Conc. Diff. = difficulty of conceptual item; Conc. Disc. = discrimination of conceptual item; Trad. Diff. = difficulty of traditional item; Trad. Disc. = discrimination of traditional item.

Table V. Item pairs with sizable performance differences

	Higher performance on conceptual items.		*Higher performance on traditional items.*	
	10% higher	*20% higher*	*10% higher*	*20% higher*
First term exam	ST2, ST3, G3, MS2, MS3, MS4, TH1	ST2, G3, MS3, MS4, TH1	P1, ST1, ST4	ST1
Second term exam	AB2, AB4	AB2, AB4	TD1, TD2, EC2, SO3,AB1, N2	TD2, SO3

Discussion and Implications

At this point, the information presented is essentially an empirical observation, predicated on the ability of the Exams Institute to organize both test construction and data collection over nationally relevant student samples. This does not imply, however, that these empirical observations provide no insight into the robustness of theories of learning or assessment related to general chemistry. In particular, it may be possible to infer some hypotheses about how the results summarized here are related to chemistry education from either a research or practice perspective.

The distinction between student performance on traditional chemistry items versus conceptual items has been a key empirical motivation for understanding how students learn chemistry for decades. One hypothesis that may be formulated from the results on this set of exams is that the work of early investigations of this phenomena (*1–8*) appears to have had an effect on instruction and student learning. Twenty years after the seminal paper from Nurrenbern and Pickering (*1*) a nationally administered exam can be crafted to measure both aspects of student learning and the resulting student performance is no longer a one-sided measure. As noted, for over 40 item pairs, performance is better on conceptual items in 19 cases and on traditional items in 21 cases. It may be that conceptual understanding gains are more substantial in material commonly covered in the first semester of general chemistry (suggested by the data in Table V), but overly broad generalizations about student conceptual understanding may be risky to make. In essence, care must be exercised to avoid having research results engender new myths about teaching and learning in general chemistry.

There are certainly several caveats that must be acknowledged relative to this conjecture. First, any assessment is the product of the efforts of the writers and carries with it the assumptions (implicit or explicit) they make in its construction. In this case, the committee that constructed the exam had the ability to look at trial test data and choose items that would allow for similar levels of performance, on average, for conceptual and traditional items. The items used in earlier research were generally designed to elicit the misunderstandings that students tend to have, so the expectations of the measurement were different. Nonetheless, it is worth noting that even the trial tests were conducted "in the wild", that is, within classroom environments where the test was part of a course.

Second, the sample of instructors who use this exam is small. The number of student performances for this sample is large, but it is possible, perhaps probable, that the instructors who choose to use this particular exam are inclined to include an emphasis of conceptual understanding in their teaching. Otherwise, they would likely use other ACS exams without this same emphasis. It may be that there are classrooms where student performance on the traditional items would be much better than conceptual, because the instructor does not emphasize conceptual understanding. Within test theory (*19*), however, this eventuality would represent a case where the test is utilized outside of its appropriate content domain. Third, general chemistry textbooks from the 1980's when the initial studies were conducted had less emphasis on particulate level, conceptual understanding of chemistry than more recent books. To enumerate this claim, counts were carried out of illustrations that depict the particulate nature of matter (PNM) in a selection of textbooks from the 1980's era, and the current era. Some judgment is required to categorize illustrations. For example, Lewis structures are considered symbolic in this context, rather than PNM illustrations. Orbital illustrations are also not included, in part because they tend to support a different form of pedagogy related to bonding rather than reactivity and in part because there has been relatively little change in the extent of these depictions utilized in texts. Illusrations are designated dynamic if they impart information about either physical or chemical change. The data from this exercise is summarized in Table VI.

The percentage of pages on which any particulate-level images are shown in the older set of books is 5.5%, while in modern texts the value is 30%. The comparison is even more dramatic when the nature of the illustration is considered. Images that imply dynamic characteristics (reactivity) at the particulate level increase from less than 1% to 5%. Dramatically, though not summarized in Table VI, in the 6 older texts, not a single end-of-chapter exercise utilizes a PNM illustration. Across the current textbooks, 236 pages in the 7 texts in the current sample contain at least one problem with such an illustration. This represents over 3% of the total pages. There seems to be little doubt that students today have a better chance of seeing PNM, conceptual depictions than students of 20 years ago.

Finally, students may be learning how to take tests that include conceptual items. From the perspective of learning theory, student test taking can often be understood in terms of which cognitive process is engaged. As categorized by Evans (*20*), there are two systems (System 1 and System 2) that humans access to accomplish a given cognitive task. System 1 tends to be more heuristic (*21*) where System 2 is more analytical and utilizes working memory (*22*). If students have been exposed to tasks that are categorized as conceptual often enough, they may have constructed useful heuristics that allow for facile answers to these questions, regardless of the putative conceptual nature of the item. In the early research in this field (*1–8*), there is little chance that the subjects had such heuristics, because the items were quite novel. Over the past two decades, more test items have emerged that are conceptual based. Student practice at answering these items increases, and test performance improves due to this practice and the heuristic reasoning it induces.

Table VI. Counts of images of the particulate nature of matter in representative general chemistry textbooks of two eras

Text (Year)	*% of pages with particulate-level images*	*% of pages with "dynamic" particulate images*
Brady and Holum, 1981	7%	1.4%
Chang, 2E, 1984	6%	0.8%
Gillespie, Humphries, Baird and Robinson, 2E, 1989	6%	0.7%
Holtzclaw and Robinson, 8E, 1988	5%	0.5%
McQuarrie and Rock, 2E, 1987	6%	1.0%
Mortimer, 6E, 1986	3%	0.7%
Average – 1980's era	**5.5%**	**0.9%**
Ebbing and Gammon, 9E	33%	6%
Gilbert, Kirss, Foster and Davies, 2E	24%	6%
Kotz, Treichel and Townsend, 7E	31%	6%
Moore, Stanitski and Jurs, 4E	25%	7%
Silberberg, 5E	35%	9%
Tro, 2E	31%	6%
Zumdahl and Zumdahl, 7E	27%	7%
Average – 2010 era	**30%**	**7%**

These, or any other conjectures or hypotheses about the observations for national samples of student performances, do not mitigate the importance of these exams as a tool for instruction and research. It does, however, point to the importance of having a theory base for both instruction and assessment in terms of using these exams. Ideally, test development "in the wild", such as that carried out by the Exams Institute will result in an instrument that has utility for research within a range of possible theory bases. The exams themselves are designed to have value for practicing educators, so long as care is taken to be sure that the content domain covered in the exam matches that of the course in which it is used. In the case of the paired-questions exams, this domain must include the relative emphasis of conceptual understanding.

Finally, it is worth noting that the sample analyzed here is large and representative of a range of instructional strategies. Even if many or most of the instructors value conceptual understanding for their students, there is little doubt that they have varying levels of emphasis on conceptual understanding. Thus, the empirical observation that an assessment can be constructed to span both conceptual and traditional domains is useful in itself. It suggests that the research results of the late 20th century, may have led to instructional changes that are having measurable improvements in student conceptual understanding in the early 21st century. The results presented here by no means prove this conjecture, but they offer tantalizing evidence that research driven curricular change can be effective.

Finally, the item analysis presented here provides an important benchmark for subsequent usage of these exams in future research. In particular, instructors who implement new pedagogies or other teaching interventions designed to enhance conceptual understanding have a well-characterized tool to measure their innovation.

Acknowledgments

The 2005 First Term General Chemistry Paired Questions and 2007 Second Term General Chemistry Paired Questions exams were created by a committee of excellent educators and researchers led by Diane M. Bunce. We gratefully acknowledge their efforts in the crafting of this exam.

References

1. Nurrenbern, S. C.; Pickering, M. *J. Chem. Educ.* **1987**, *64*, 508–510.
2. Pickering, M. *J. Chem. Educ.* **1990**, *67*, 254–255.
3. Sawrey, B. A. *J. Chem. Educ.* **1990**, *67*, 253–254.
4. Nakhleh, M. B. *J. Chem. Educ.* **1992**, *69*, 191–196.
5. Nakhleh, M. B. *J. Chem. Educ.* **1993**, *70*, 52–55.
6. Nakhleh, M. B.; Mitchell, R. C. *J. Chem. Educ.* **1993**, *70*, 190–192.
7. Zoller, U.; Lubezky, A; Nakhleh, M. B.; Tessler, B.; Dori, Y. J. *J. Chem. Educ.* **1995**, *72*, 987–989.
8. Nakhleh, M. B.; Lowrey, K. A.; Mitchell, R. C. *J. Chem. Educ.* **1996**, *73*, 758–762.
9. Niaz, M. *Int. J. Sci. Educ.* **1995**, *17*, 343–355.
10. Cracolice, M. S.; Deming, J. C.; Ehler, B. *J. Chem. Educ.* **2008**, *85*, 873–878.
11. Eubanks, E. D.; Eubanks, L. P. *General Chemistry Special Exam, First Term*; American Chemical Society, Division of Chemical Education, Examinations Institute: Washington, DC, 1997.
12. Holme, T. A.; Murphy, K. L. *General Chemistry First Term Paired Questions Exam*; American Chemical Society, Division of Chemical Education, Examinations Institute: Washington, DC, 2005.
13. Holme, T. A.; Murphy, K. L. *General Chemistry Second Term Paired Questions Exam*; American Chemical Society, Division of Chemical Education, Examinations Institute: Washington, DC, 2007.
14. Smith, K. J.; Metz, P. A. *J. Chem. Educ.* **1996**, *73*, 233–235.
15. Sanger, M. J.; Greenbowe, T. J *J. Res. Sci. Teach.* **1997**, *34*, 377–398.
16. Raviolo, A. *J. Chem. Educ.* **2001**, *78*, 629–631.
17. Holme, T. A. *J. Chem. Educ.* **2003**, *80*, 594–598.
18. Holme, T. A; Murphy, K. L. *J. Chem. Educ.* **2011**, *88*, 1217–1222.

19. Crocker, L.; Algira, J. *Introduction to Classical and Modern Test Theory*; Holt, Reinhart and Wilson: New York, 1986.
20. Evans, J. S. T. B. *Ann. Rev. Psychol.* **2008**, *59*, 255–278.
21. Chen, S.; Chaiken, S. *Dual-Process Theories in Social Psychology*; Chaiken, S., Trope, Y., Ed.; Guilford: New York, 1999; pp 73–96.
22. Baddeley, A. D.; Hitch, G. J. *The Psychology of Learning and Motivation*; Bower, G. A., Ed.; Academic Press: New York, 1974; Vol. 8, pp 47–90.

Chapter 13

Diverse Methodologies Used To Challenge Myths

Jessica R. VandenPlas*,[1] and Diane M. Bunce[2]

[1]Chemistry Department, Grand Valley State University, Allendale, Michigan 49401
[2]Chemistry Department, The Catholic University of America, Washington, DC 20064
***E-mail: vandenpj@gvsu.edu**

The chapters in this book present a wide range of research questions and methodologies challenging commonly held myths of teaching and learning. There are many valid research designs that can be used to conduct quality research including quantitative methods, qualitative methods, test analysis theories, and in many cases utilizing multiple experiments to address complicated questions. Each chapter in this book can be viewed independently as an example of a typical type of research. This final chapter will revisit the research presented previously and discuss it in terms of the research methodologies used. The purpose is to help the reader better see the infrastructure of the research presented and how the methodology employed matches the research questions asked.

Introduction

One of the purposes of this book was to provide reports of research that addressed some popular myths about teaching and learning. The questions investigated included the following: How long do students pay attention in lecture? How long do they retain knowledge after a test?; Can students do inquiry even if their teachers don't think they can? Do students do better in semester-long or intensive summer school chemistry courses? Do student-prepared formula sheets help or hinder students' achievement on tests compared to formula sheets prepared by teachers? Is the same amount of content taught in an activity-based vs. a lecture-based course? Can we analyze test questions to help identify topics that are difficult for students?; Can standardized tests be used to gauge whether we as a field have made progress teaching conceptual understanding? What are effective ways to choose and implement different visualization methods in chemistry? What is the process of developing valid and reliable assessments? How can surveys be used to track changes in student use of curriculum options over time?

This is a diverse list of questions that are tied to myths (cultural beliefs) that research can investigate and challenge. The take home message of the book is that any belief that appears to explain some aspect of the teaching/learning process is worth investigating in an attempt to place it on a firmer footing than hearsay alone. The goal is to put the process of teaching and learning on a firm foundation supported by research results. This is easier said than done. Teaching and learning are complicated processes that encompass a large number of variables, only some of which we may be aware of. These variables may include student aptitude; student self confidence in pursuing understanding of difficult topics; time available to learn the material; the way the material is presented in terms of visual, audio or kinesthetic approaches; the framework of the course and the materials provided to support learning; the empathy and support provided by the teacher; and the appropriateness and accessibility of testing measures. In some situations, we may not even be aware of what the variables are as well as not knowing how to adequately measure or control them.

In reading the research presented in this book, we can analyze both the types of questions asked and the research methodologies employed. No two studies reported here are alike. They all ask different questions and use different methodologies that help address these unique questions. Rather than just looking at the results of these studies, this chapter will analyze the methodologies used to address the questions raised by the researchers. The purpose of this chapter, then, is different from the rest of the book. Here we are more interested in *how* the questions were addressed rather than *what* the results of the research were. To this end, we will revisit the chapters analyzing the research methodologies employed and how these methodologies were used to address the questions asked.

Quantitative, Statistical Research Designs

One of the most common research designs, shared by several studies in this book, involves the collection of quantitative data that can be analyzed using inferential statistics (*1*). Quantitative data can be collected in many ways, from traditional in-class quizzes and exams, to novel methods, such as the use of clickers described in the chapter by Neiles et al. (Chapter 5). Before any quantitative data is collected, researchers have many decisions to make in how to process and analyze the data to best address the research question being asked. One of the most common ways of dealing with quantitative data is to use significance testing to identify differences, or the lack thereof, among datasets. Such methods were used to address several research questions in this book including topics such as student achievement, course design, and attention allocation. Each study applied these methods in a unique manner that was determined by the question that was being asked.

Studies Utilizing Quantitative Methods

The study by Bunce and VandenPlas (Chapter 2) was conducted to investigate how student knowledge changed over time. In order to address such a research question, the researchers must first operationally define terms such as "knowledge" for the purposes of the study. In this case, the researchers used student performance on in-class quizzes and exams as a way of approximating student knowledge. The quiz and exam questions used for the study were short answer questions, providing students' free responses as data for the investigation. These data could be analyzed several ways, ranging from a direct qualitative analysis of the language used in the answers, to the more quantitative methods analyzing achievement employed in this study. In this case, the student answers were rubric-scored by the researchers, a common method of converting free responses (qualitative data) into numerical (quantitative) data for the purpose of statistical analysis. This is one way to simplify a large data set for a more manageable analysis. Here it also allowed the researchers to compare student scores across testing occasions in a statistical manner. This provided a clearer picture of the changes that occurred and how these were influenced by variables such as course type and students' logical reasoning ability.

Because these data were analyzed in a quantitative manner, the researchers were able to document the trends observed in student scores (no significant decay for nursing and high school students, but immediate decay for nonscience majors). Although the trends could be identified, the *reasons* for these trends could not. This would require asking a different research question: *why* student knowledge changed over time rather than *whether* it changed over time. Clearly, one must document *if* a change occurs before asking *why* it occurs. Collecting student quiz/exam responses was a reasonable method of determining whether students' knowledge changed over time, but additional data would be needed to answer the question of why this happened in some classes and not others.

As a start towards addressing competing explanations for the resulting data, the researchers collected and analyzed additional qualitative data in the form of student evaluations. This was used to investigate the hypothesis that student motivation may play a role in how knowledge changes over time. The evaluation did help to eliminate some possible explanations for why the data provided the trends it did. Because different measurement tools were used, the study by Bunce and VandenPlas is considered a mixed methods design (*2*), utilizing both quantitative analyses (statistical analyses of student quiz and exam scores) and qualitative analyses (rubric-scoring student free responses and a survey to collect data on student motivation). Mixed methods designs allow for triangulation of data. Data triangulation is the process of collecting data from more than one source or instrument and analyzing it to reveal inconsistencies, identify contradictions, or to provide complementary data to support interpretation (*3*). Triangulation also helps to control overall threats to validity by balancing the individual threats of different data collection methods. Mixed methods designs can therefore often provide a more complete picture of the research questions posed, and the soundness of the data interpretation by the researchers.

In the study conducted by VandenPlas et al. (Chapter 3), the authors also addressed a research question regarding student achievement. In this case, the authors had competing hypotheses regarding how students would perform in both the short-term and long-term after using formula sheets on examinations. Both competing hypotheses predicted students would do better on examinations when they had prepared their own formula sheets, either because students had spent extra time organizing their knowledge or because the formula sheet served as an external storage device. The research hypotheses differed, however, on the long-term effects of allowing students to use their own formula sheets. The overarching question in this study was: would students learn more overall because of the time spent organizing their sheets and identifying important concepts, or would they learn less because they had these sheets to rely on? To test these competing hypotheses a two-part study was devised.

As in most achievement studies, control of variables was important. Because formula sheet use on a single exam was hypothesized to affect a student's knowledge structure on the topics covered by that exam, and due to the integrated nature of chemistry concepts, students were assigned to one research treatment for the entire semester. Course sections were randomly designated as "experimental" and "control" groups, and all students enrolled in those sections experienced the same presentation of material and exam questions. In the "experimental" section, students were allowed to prepare their own formula sheets, whereas in the "control" section students were given the standard instructor-prepared formula sheets. Confounding effects can arise when entire sections are sampled in this way, including self-selection into certain class sections, time-of-day effects, and other variables. To control for these possible effects, the researchers had to demonstrate that the sections being compared were equivalent prior to beginning the study. It is common to make this comparison statistically on some pertinent achievement measure or demographic prior to the research intervention. In this study, the researchers used student achievement without formula sheets on several quizzes prior to the start of the research to demonstrate that the sections performed equivalently under similar test conditions.

The study by VandenPlas et al. is an example of a quantitative study, relying only on quantitative data and statistical analyses to address the research question. Although the results of the statistical tests for this study were all non-significant, showing that the intervention (student-prepared formula sheets) did not affect student performance and thus failing to confirm the initial research hypotheses, the outcome did address the question asked. In this study, the authors set out to *prove* a common classroom myth regarding how formula sheets hinder student learning. In the end the non-significant statistical results helped to *disprove* the myth instead.

The study by Hall et al. (Chapter 8) is another example of a quantitative study based on research questions regarding student achievement. In this study, the authors compared chemistry content knowledge between students taking chemistry courses during a 3-week summer session with those in a traditional 15-week semester. In addition, chemistry content knowledge of students in these two course types was compared by life experience (time since high school graduation). As previously discussed, the authors first had to define "chemistry content knowledge" in order to devise a method of addressing their research questions. In this case, student achievement on an in-class exam was used as a measure of chemistry content knowledge. This provided quantitative data, and allowed for a statistical comparison between groups of students.

A unique aspect of this study is that instead of collecting a new set of data to address their research question, the authors conducted a secondary analysis of existing longitudinal data. A secondary analysis is an examination of data that goes beyond what was originally intended when the data were collected. The researchers discuss the pros and cons of using a secondary analysis in their chapter, but the main advantage is clear: data *already existed* that would allow the researchers to investigate their research questions. The data only needed to be analyzed in a different way. This is a good use of longitudinal data, which is costly both in terms of time and money to collect. This study utilized four years worth of data, which would have required a significant amount of effort for the researchers to collect. Instead, the researchers were able to utilize data that had already been collected for a different purpose. Although this gave the researchers less freedom to design the study, the tradeoff in the amount of data collected was deemed worthwhile.

Results of this study showed significant achievement differences between students in the two courses. The statistical results demonstrate that students enrolled in the 3-week summer session gained more chemistry content knowledge than those in the traditional 15-week semester. Students who had graduated from high school more than 5 years prior to taking the course were also shown to possess more chemistry content knowledge than those who graduated less than 5 years prior to taking the course. As in the Bunce and VandenPlas (Chapter 2) study, the existing data could not address *why* these differences occurred, but the data did allow the researchers to address their initial research questions regarding *if* the differences existed. The authors suggest that a mixed-methods study, using qualitative data such as interviews in addition to the quantitative analysis of student performance, would have been a good methodology to address these additional research questions. Using existing data restricted this study to a quantitative design.

The article by Neiles et al. (Chapter 5) is an example of a project devoted not to chemistry content, but to student behavior. This study sought to identify the pattern of student attention during class specifically, at what point attention lapses occurred and the length of these lapses. Because the research question dealt with timing issues, it was important for the researchers to devise a real-time data collection method. While the researchers could have collected retrospective reports from students after class or as part of a daily or weekly survey, this would not have adequately addressed the time-sensitive nature of the research questions, such as whether attention remained constant during lecture, or varied based on teaching pedagogy. Instead, the researchers used a novel data collection method—utilizing student response systems ("clickers") to collect student self-reports of attention lapses in real-time during class.

The data collected in this study is qualitative in nature (student self-reports of attention lapses), but the use of clickers converts this to quantitative, numerical data (number of lapses, the time each lapse occurred, and student-report of lapse length). From there, the authors use statistical methods to analyze the data specifically, significance testing used to compare the occurrence of attention lapses throughout the courses studied. This use of significance testing methods shows that it is not limited to achievement data or test scores as described in previously discussed studies. The same statistical methods can be applied to any set of quantitative data as in this case where behavioral data reported by students themselves was used.

In this study, two of the three courses investigated showed nonsignificant results. The third course showed more variation in attention lapses reported by students. This demonstrates the importance of gathering a robust data set, such as the three courses, rather than just one, used in this study. The courses were analyzed separately due to the inherent differences among them, but these separate analyses also allowed the researchers a finer level of investigation. Here the significant results found for a single course may have been lost if data from these three disparate courses had been aggregated.

Qualitative Research Designs

In many cases, quantitative data is not appropriate for addressing a given research question. Researchers sometimes seek to investigate research questions describing a particular classroom situation, discussing *why* a particular effect occurs, or the process by which educational changes occur. To answer questions like these, qualitative data is often a more suitable and powerful tool. Qualitative data can take the form of interviews, surveys, observations, document analysis, or field notes (*3*). Rather than using inferential statistics, qualitative coding and descriptive statistics (describing the frequency of certain occurrences, for example) are more appropriate methods of analyzing qualitative data. Many studies described in this book include at least some qualitative methods to aid in data triangulation resulting in mixed-methods designs. The studies described here, however, focus exclusively on qualitative designs.

Studies Utilizing Qualitative Methods

The chapter by Daubenmire et al. (Chapter 7) explored two research questions regarding whether a curricular intervention would result in high school chemistry teachers changing their instructional practices to include more inquiry activities. If so, then a second question concerning if the teachers would perceive shifts in student abilities to perform such inquiry activities was asked. These research questions focus not on student knowledge acquisition itself, but rather on classroom practices and how to bring about change in such practices. To answer these questions, the researchers used qualitative research methods involving the practices of teachers and students at multiple institutions.

The first method used to gather data in this study was the use of focus groups. Conducting focus groups at various points throughout the study enabled the researchers to gather feedback from multiple teachers simultaneously, without having to interview each teacher individually. The researchers then qualitatively coded the teacher feedback using a theoretical framework appropriate for their investigation called SWOO (Strengths, Weaknesses, Obstacles, and Opportunities of program implementation). This data provided evidence that teachers were indeed moving in the direction of adopting more inquiry in the classroom as a result of the support provided within the teacher training and curricular initiative.

To triangulate this data, the researchers used two surveys of the teachers themselves. The first was a self-assessment developed specifically for this project, asking teachers to rank their own progress and that of their students towards achieving specific program benchmarks (including conducting inquiry activities). The second was an existing survey, called the Survey of Enacted Curriculum (SEC), which was given to both teachers taking part in the curricular and training intervention being studied and those who were not. This survey asked teachers to identify the time they spent in various classroom activities, including focusing on specific concept or skill developments. Data from both of these surveys agreed with the earlier focus group data, and showed that teachers taking part in the curricular intervention spent more time on average using inquiry activities or developing inquiry skills than those who were not part of the initiative.

Finally, to further investigate teacher and student use of inquiry, researchers analyzed reports made by program coaches from the training initiative. The program coaches worked with teachers one-on-one on a regular basis to help them adopt the new curricular approach and to mentor them in the classroom. The coaches provided regular reports to the researchers, and these reports were analyzed for both the occurrence and frequency of representative quotes, which the researchers used in their analysis. This study is a good example of a qualitative research design, but perhaps more importantly, shows the importance of data triangulation. Although the researchers could have collected only one type of data (focus group feedback, for example), they collected multiple data sets from multiple sources, including teachers involved in the study, teachers outside the study, and program coaches to inform their conclusions. This provided a rich dataset that the authors used to directly address their research questions. The end result demonstrated that teachers (and their students) exposed to the particular curricular and teacher training intervention were, in fact, successful in moving towards more inquiry based activities in the classroom.

The study by Pienta (Chapter 9) is another example of a study that focused on non-content material related to chemistry, and which used qualitative methods, in this case, surveying. In this study, the author was interested in how students valued various course components, such as lecture or textbooks, in their learning. This study also serves as an example of longitudinal research, in that data was collected for 10 years using a single instrument. This allowed the researcher to identify trends in student responses and how these trends changed over time. While data from a single data collection may sometimes be skewed by mitigating factors, a longitudinal study provides a robust dataset for analysis and increases confidence in the interpretation of the results of the analyses by the author.

Pienta's research used a Likert-scale survey to gather information from students regarding self-reports of the relative value and time spent interacting with each course component. Because of the nature of the research questions being investigated, inferential statistical analyses and significance testing were not appropriate for this study, even though the qualitative data collection method (surveying) used provided numerical data (Likert scores). Instead, descriptive statistics, including frequency counts of student responses in each Likert category, are reported and graphed, allowing the researchers to identify the students' value of each component. Changes in these trends over time were also investigated. Additional descriptive statistics such as mean and standard deviation were reported for student self-reports of time spent on various course components and non-course related activities.

Using Test Analysis Theories To Address Research Questions

One thing that many of the previously discussed studies have in common is the use of an instrument, whether quantitative or qualitative, to collect data. When chemists hear the word "instrument" they no doubt think of an HPLC or GC-MS, but for chemical educators, "instruments" are data collection tools such as attitudinal surveys and chemistry content exams. While many of the studies discussed in this book make use of existing instruments, such as the GALT or TOLT exams, several also develop their own exams or surveys. Although the selection, development, and validation of these instruments are not discussed in depth in these individual studies, the chapter by Barbera and VandenPlas (Chapter 11) provides a discussion of the considerations that should be made when conducting research that utilizes such instruments. The authors discuss existing literature on how to develop valid and reliable instruments for use in research, including how existing instruments may need to be evaluated before use in a study. For further discussion on designing tests and surveys, consult Scantlebury and Boone's chapter (*4*) in a previous symposium series volume.

Once a robust instrument has been built or selected, there are many statistical tools available to analyze the data collected. Classical Test Theory (CTT) and Item Response Theory (IRT) are two frameworks that can be used to analyze data collected from an instrument, and two studies in this book, utilizing these models are discussed below.

Classical Test Theory

The study by Holme and Murphy (Chapter 12) is an example of research utilizing Classical Test Theory (CTT) to analyze student performance data on a developed instrument. In this study, the researchers looked at a large data set collected from students who had taken the American Chemical Society's Paired-Questions exams (first- and second-semester) at different institutions. In this case, item level statistics were used not only to develop the instrument itself (which was not the focus of this particular study), but also to separately investigate a research question regarding student achievement on two different types of chemistry content questions, namely, conceptual and algorithmic questions.

Although the researchers discuss how item statistics from CTT, such as difficulty and discrimination, were used to select the items that ultimately became part of the final instruments, the more interesting use of these statistics comes in analyzing student performance on the final instruments themselves. Item difficulty, in particular, is used to compare traditional and conceptual questions on a single topic to identify where students have difficulty. The item level statistics provide a picture of the general trends in student achievement for both conceptual and algorithmic questions.

This study is an example in which overall student achievement on the exam, as used in some of the studies discussed previously, could not directly address the proposed research questions. Even pooling all traditional questions and comparing them to the pool of all conceptual questions would have lost the fine-grained nature of this analysis, and provided an unconvincing answer to the research questions presented in the study. Analyzing individual item pairs allowed the researchers to compare traditional versus conceptual questions for individual topics in chemistry, and to tease out which topics were more difficult for students. The authors used these results to address the question of whether students were more likely to successfully answer traditional or conceptual questions today, as compared to results from earlier research.

Item Response Theory

The research presented by Shurmeier et al. (Chapter 10) is another example of analyzing a large dataset collected over many years on a large number of students. In this study Shurmeier et al. used Item Response Theory (IRT) in a manner similar to that of Holme and Murphy's use of CTT as discussed previously (Chapter 12), in order to answer research questions regarding student achievement and capabilities in chemistry.

IRT, like CTT, can be used to determine item difficulties and discriminations, but IRT can also assign students to individual ability levels. The authors indicate that results from early IRT analyses were used to eliminate poorly performing questions from the final instrument and to measure test reliability. Here, the statistics were used to directly answers the main research question, namely, are some topics in chemistry more difficult for students than others?

Using IRT, the authors were able to identify topics that were consistently easy for students, such as unit conversion and balancing equations, and those that were consistently difficult, such as the particulate nature of matter and intermolecular forces. The analysis also allowed the researchers to analyze students' performance by their ability, and to predict which topics would be difficult for different ability-level students. This reflects a fine grain of detail, and allows the research to identify the best questions for sorting students by letter grade (A, B, C, D, and F). For example, questions dealing with the particulate nature of matter were routinely missed by students who received a low C, D, or F on the exam, while A, B, and high C students did not demonstrate difficulty with these questions. This same analysis was carried out for other types of questions and in this way, it was possible to identify topics that were both difficult for *all* students enrolled in the course and which were difficult for only lower performing students.

To address this research question, Shurmeier et al. needed a research design that would allow them to compare student achievement on questions across a large number of topics. In this case, achievement was defined as student performance on course exams, which allowed for multiple questions on each topic to be tested. Data were collected on multiple versions of the in-class exams over several years, using several thousand students. While there may be other research methodologies capable of addressing the research question in this study, the large amount of data collected was handled well by IRT thus providing a robust statistical analysis.

Assembling a Body of Literature To Address Complex Research Questions

Occasionally we are faced with a research question for which a large body of research already exists. A thorough analysis of the literature is a time consuming process, but synthesizing existing work can frequently provide satisfying results. In situations where individual research studies each add a piece to the puzzle, it may take several years of research, synthesized into a single body of work, to address more complex research questions. Two chapters in this book provide discussions of how the researchers have compiled a large body of work through their own research, and that of others, to best address long-standing myths from multiple angles.

Integrating Multiple Experiments

The chapter by Oliver-Hoyo (Chapter 4) is one such example. In this case, the author presents a review of her own research, consisting of four separate studies on the use of a new curriculum entitled $cAcL_2$ (concept Advancement through chemistry Lab-Lecture). The benefit of a review of the literature in this case is that it allows the data from multiple independent studies to be compared simultaneously and considered as a whole. This provides triangulation for the individual datasets, and offers compelling evidence to address the research question of how this particular curricular intervention (a lab-lecture hybrid) compares to the effectiveness of covering content with traditional lecture-based methods.

This chapter discusses four separate studies that have previously been published, and which provide evidence for the efficacy of the $cAcL_2$ method. The first study is an achievement study, comparing exam scores between students experiencing the $cAcL_2$ method and those in a traditional lecture class. This was a quantitative design, in which exam scores were compared using statistical significance tests. The results of these tests suggested that the $cAcL_2$ students performed significantly better than the traditional students on many exams.

Two attitudinal studies were also conducted, the first using surveys developed specifically for this project, and the second using an array of qualitative data collection methods including interviews, surveys, student self-reports via journaling, and field note entries from the researchers themselves. Although these qualitative data on student attitudes could not address how the $cAcL_2$ method influenced students' understanding of chemistry, they were used to support the observation that students viewed the $cAcL_2$ method more positively as the semester progressed.

Finally, a study was described in which the researcher studied students' higher order cognitive skill development. To do this, qualitative data were collected in the form of student reflections on their own problem solving methods after completing several conceptual chemistry problems. The problem solutions themselves were not studied. Instead, the retrospective reflection on the *methods* the student used to solve the problems was qualitatively coded by the researcher using the theoretical framework of Bloom's taxonomy. These codes were compared over time resulting in the emergence of a general trend in which students exposed to the $cAcL_2$ method showed increased use of higher order (abstract) reasoning to solve problems.

Taken together, these four studies provide a well-rounded view of the effects of using the $cAcL_2$ method in the classroom. While each study individually is a crucial piece to the puzzle, combining multiple studies in this way provides a much clearer overall picture of the body of research used to address the over arching research question.

Addressing Pervasive Myths

Williamson (Chapter 6) provides another review of the literature, including many of her own studies, in her chapter discussing the use of visualizations in the chemistry classroom. This review aggregates over a decade's worth of research on the topic, and uses this existing data to challenge several myths about the use of visualizations, such as animations, in chemistry.

This chapter demonstrates how pervasive myths can be in the area of teaching and learning despite existing proof to the contrary. For example, the first myth addressed by Williamson centers on how much time must be sacrificed from lecture in order to use animations. To counter this myth, Williamson cites her own 1995 study, in which students' conceptual understanding of the particulate nature of matter was compared after viewing animations for varying amounts of time. This early study provides compelling evidence that viewing animations does not require sacrificing large amounts of class time to reap significant benefits. Yet this study completed over a decade ago and providing solid statistical evidence to demonstrate this point, has not been sufficient to completely put this particular myth to rest.

Williamson discusses six myths in total, each with a body of literature that should effectively challenge the myth. The fact that these myths are so pervasive, even in the face of such compelling research, is a testament to the power that myths have within the culture of education. This chapter shows that simply conducting and publishing good research is not enough. Researchers must advocate against myths and continue to work to educate chemical educators of the research done to disprove them.

The Williamson chapter is also a good example of the type of literature search that should be undertaken in the process of forming one's own research questions.

Producing a review of the literature of this scope allows one to identify common theoretical frameworks in the field, holes in the existing research, and prevents the researcher from reinventing the wheel. This type of review aids in the development of robust research questions that can serve as the starting point for quality research. Such research not only challenges common myths, but adds value to the body of literature as well.

Conclusions

This book encompasses a wide variety of research questions. The researchers who addressed these questions each chose a different methodology including quantitative, qualitative and mixed methods designs. Some researchers chose questions that rely heavily on statistical analysis to answer. while others used the wealth of information in the literature to help challenge myths. Each of these methods provides different advantages and disadvantages. The researcher's role is to match the questions asked to the methodology that best addresses those questions. Since myths are tied to beliefs about teaching and learning, myths are pervasive in the culture of education and are difficult to isolate and confront. It requires carefully crafted research questions and well designed studies to confront these myths. Only with the data from thoughtful experiments, such as those presented in this book, can teaching and learning myths be effectively challenged.

References

1. Sanger, M. J. Using Inferential Statistics to Answer Quantitative Chemical Education Research Questions. In *Nuts and Bolts of Chemical Education Research*; Bunce, D. M., Cole, R. S., Eds.; ACS Symposium Series 976; American Chemical Society: Washington, DC, 2007; pp 101−133.
2. Towns, M. H. Mixed Methods Designs in Chemical Education Research. In *Nuts and Bolts of Chemical Education Research*; Bunce, D. M., Cole, R. S., Eds.; ACS Symposium Series 976; American Chemical Society: Washington, DC, 2007; pp 135−148.
3. Bretz, S. L. Qualitative Research Designs in Chemistry Education Research. In *Nuts and Bolts of Chemical Education Research*; Bunce, D. M., Cole, R. S., Eds.; ACS Symposium Series 976; American Chemical Society: Washington, DC, 2007; pp 79−99.
4. Scantlebury, K.; Boone, W. J. Designing Tests and Surveys for Chemical Education Research. In *Nuts and Bolts of Chemical Education Research*; Bunce, D. M., Cole, R. S., Eds.; ACS Symposium Series 976; American Chemical Society: Washington, DC, 2007; pp 149−169.

Editor's Biography

Diane M. Bunce

Diane M. Bunce is a chemistry professor at The Catholic University of America in Washington, DC, where she has taught since 1985. She is the co-editor of *Nuts and Bolts of Chemical Education Research*, ACS Symposium Series 976, and the Associate Editor of the *Journal of Chemical Education*, Chemical Education Research Section. Diane served as an original author on all three of the American Chemical Society's curriculum projects: *ChemCom: Chemistry in the Community*, *Chemistry in Context*, and *Chemistry*. Her research focuses on how students learn chemistry and the mismatch between the way we teach and how the brain operates.

Indexes

Author Index

Subject Index

D

E

F

J

K

L

M

N

O

P

Q

R

S

T

U

V

W

X

Z